# AI
# in Civil Engineering
# and Construction

To the point handbook on Civil Engineering and Construction with AI

By

Dr. Herald Noronha, PMP
Dr. Raviraj Jain
Dr. Prakash Kumar Udupi

Yashas Publications

# Preface

Artificial Intelligence (AI) plays a significant role in the dynamic landscape of civil engineering and construction. AI integration in civil engineering becomes necessary to meet the evolving challenges and technological demands of 21st century. AI making inroads in the field of civil engineering and construction as a revolutionary force. AI enables integration of cutting-edge technologies and reshaping the way we plan, design and construct the infrastructures that define our immediate societies.

This book titled AI in Civil Engineering and Construction facilitate the readers to explore the intersection of artificial intelligence and the built environment by offering a comprehensive knowledge with real-world examples and use cases on how AI is revolutionizing every aspect of civil engineering and construction projects. Right from the conceptualization to completion, in every area of projects, AI is altering the traditional paradigms by introducing unprecedented efficiency, precision and sustainability into the construction industry.

This book also navigates the readers covering wide areas of AI applications in civil engineering and constructions ranging from intelligent design and optimization algorithm to construction site automation and predictive maintenance. The book further enables the readers to embark on a journey, that unlocks the transformative potential of AI technologies by providing a roadmap for the future of AI integration in civil engineering and construction.

| Sl. No | Topics | | Page No. |
|---|---|---|---|
| **1.** | **Introduction to Artificial Intelligence in Civil Engineering** | | **1** |

## 3. Basics of Artificial Intelligence 54

# 1. Introduction to Artificial Intelligence in Civil Engineering

Artificial Intelligence (AI) has emerged as a transformative force in various industries, and civil engineering is no exception.

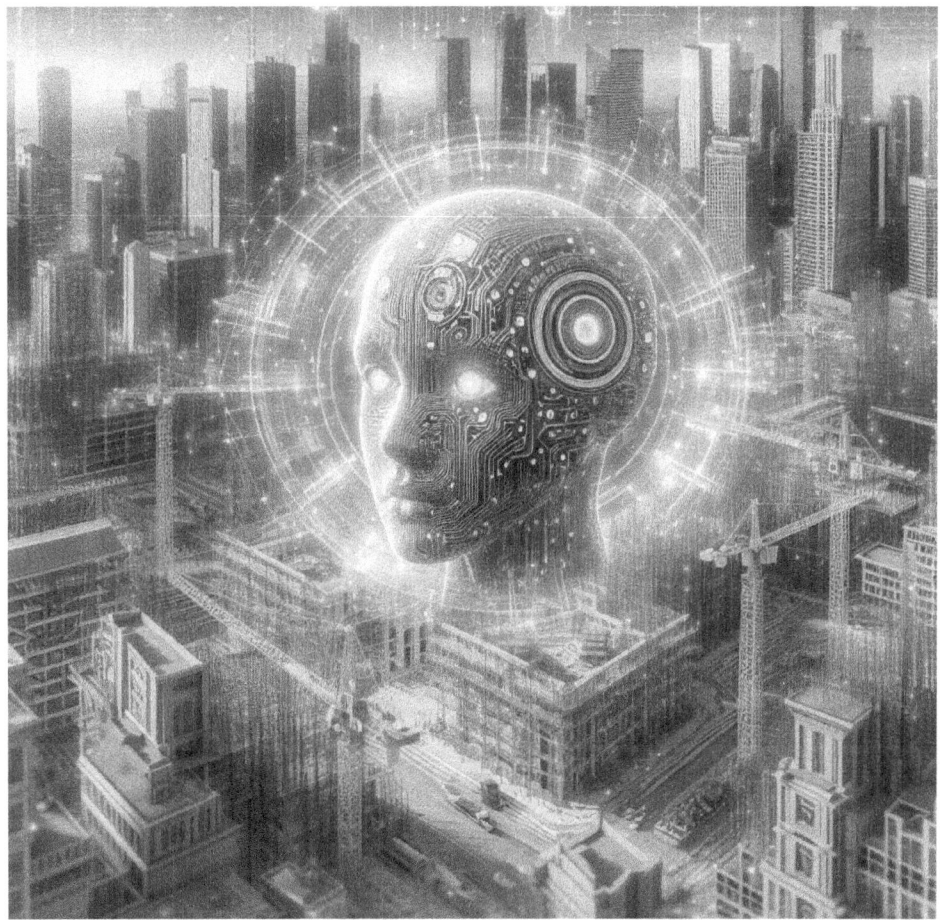

In the realm of infrastructure development and construction management, AI presents a paradigm shift, reshaping traditional practices and introducing novel approaches to problem-solving. At its core, AI encompasses a suite of technologies and algorithms designed to simulate human intelligence, enabling machines to process information, learn from experiences, and make informed decisions. This introduction aims to explore the profound implications of integrating AI into the field of civil engineering, acknowledging both its potential benefits and the challenges that come with such technological advancements.

In recent years, the integration of AI into civil engineering practices has gained momentum, driven by the increasing complexity of projects and the need for efficient, data-driven solutions. This shift represents a departure from conventional methods, offering a more proactive and adaptive approach to project planning, design, and execution. One of the key aspects of AI in civil engineering lies in its ability to analyze vast amounts of data swiftly and accurately. This analytical prowess extends beyond mere data processing; it enables the identification of patterns, trends, and potential risks that might otherwise elude human observation.

AI's influence in civil engineering manifests in diverse applications, ranging from project planning and resource allocation to design optimization and structural health monitoring. Through machine learning algorithms and predictive analytics, AI empowers engineers to make informed decisions, enhancing the overall efficiency and sustainability of construction projects. Furthermore, AI-driven technologies, such as robotics and autonomous vehicles, are revolutionizing on-site operations, promoting safety, precision, and cost-effectiveness.

As the industry adapts to these technological advancements, it is essential to address ethical considerations and potential challenges associated with the widespread implementation of AI. Issues related to bias in algorithms, data privacy, and the displacement of traditional job roles require careful scrutiny. Balancing the potential benefits of AI with ethical considerations becomes paramount in ensuring the responsible and equitable integration of these technologies into civil engineering practices.

As we delve further into the transformative landscape of AI in civil engineering, it becomes evident that the implementation of these technologies extends beyond mere optimization. The convergence of AI with other cutting-edge technologies, such as virtual reality and augmented reality, introduces a new dimension to project design and visualization. Engineers can now leverage immersive technologies to create realistic simulations, fostering better communication and understanding among project stakeholders.

Additionally, the role of AI in sustainable practices within civil engineering cannot be overstated. Through intricate analyses of environmental data and the integration of energy-efficient solutions, AI contributes to the development of eco-friendly infrastructure. This commitment to sustainability aligns with global initiatives for greener construction practices, demonstrating the potential of AI not only as a technological enabler but as a catalyst for positive environmental impact.

In the context of project monitoring and control, AI's real-time capabilities empower project managers with dynamic insights. By continuously analyzing project data, AI facilitates the identification of potential bottlenecks, risks, and deviations from the planned schedule. This proactive approach enables timely interventions, ensuring projects stay on track and within budget.

As AI continues to evolve, its influence extends to the workforce, necessitating a shift in skillsets and fostering a culture of continuous learning. The collaboration between humans and machines, often referred to as augmented intelligence, emphasizes the synergy between human intuition and AI's analytical capabilities. This collaborative approach not only enhances project outcomes but also underscores the importance of adaptability in the face of technological advancements.

Despite the remarkable strides made by AI in civil engineering, there are challenges that demand careful consideration. Ensuring data security, mitigating biases in algorithms, and addressing concerns related to job displacement require a comprehensive approach. Ethical frameworks must be established to guide the responsible deployment of AI, safeguarding against unintended consequences and promoting transparency in decision-making processes.

The integration of artificial intelligence (AI) in civil engineering and construction involves a systematic approach comprising nine interconnected stages. Beginning with data collection from diverse sources like sensors and historical project data, AI plays a pivotal role in processing large datasets, uncovering patterns, and aiding in comprehensive analysis. Subsequent stages include AI-driven data preprocessing for accuracy, predictive analytics for forecasting project outcomes, structural health monitoring for real-time anomaly detection, and AI-optimized project planning and scheduling. The utilization of AI extends to autonomous equipment and robotics, enhancing navigation and adaptability on construction sites. Supply chain optimization, quality control, and safety monitoring benefit from AI's predictive capabilities, while post-construction monitoring leverages AI for assessing long-term structural performance. The following sequential process illustrates implementation of AI in civil engineering and construction.

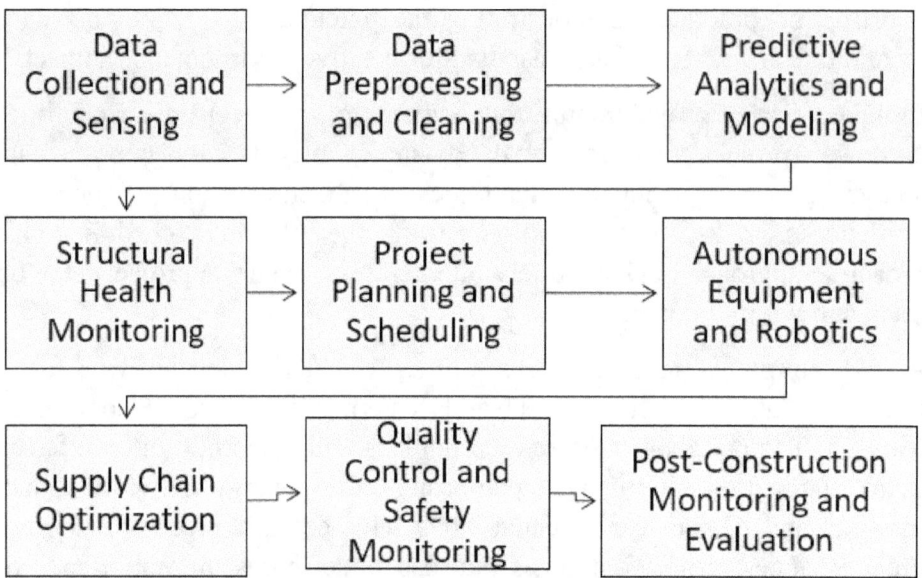

This sequential process demonstrates how AI can be integrated into various stages of civil engineering and construction, improving efficiency, accuracy, and overall project success. Keep in mind that the implementation may vary based on specific project requirements and goals.

Looking ahead, the future of AI in civil engineering holds exciting possibilities. The exploration of novel materials, the development of smart cities, and the integration of AI with emerging technologies are poised to redefine the industry further. These advancements underscore the need for ongoing research, collaboration, and a commitment to ethical best practices to harness the full potential of AI in shaping the future of civil engineering and construction management.

In conclusion, the integration of AI in civil engineering represents a groundbreaking evolution in the industry's capabilities. From streamlining project workflows to enhancing decision-making processes, AI holds the potential to usher in a new era of efficiency and innovation. However, careful consideration of ethical implications and the proactive management of challenges is crucial to ensuring a harmonious coexistence between human expertise and artificial intelligence in the field of civil engineering. The subsequent chapters will delve deeper into specific applications, challenges, and future prospects of AI in the realm of civil engineering and construction management.

## 1.1 Overview of Artificial Intelligence

Artificial Intelligence (AI) represents a transformative branch of computer science dedicated to creating intelligent agents capable of mimicking human-like cognitive functions.

At its core, AI seeks to develop algorithms, systems, and technologies that can perceive, reason, learn, and adapt to complex tasks. The overarching goal is to enable machines to perform tasks that traditionally require human intelligence, opening new frontiers in problem-solving, decision-making, and automation.

The foundation of AI lies in its ability to process and analyze vast amounts of data. Machine learning, a subset of AI, empowers systems to recognize patterns within data and improve their performance over time without explicit programming. This adaptability is a key feature that distinguishes AI from conventional software, allowing it to evolve and optimize its performance based on experience.

A comprehensive overview of artificial intelligence (AI) shown below spans key aspects vital to understanding its role in industries.

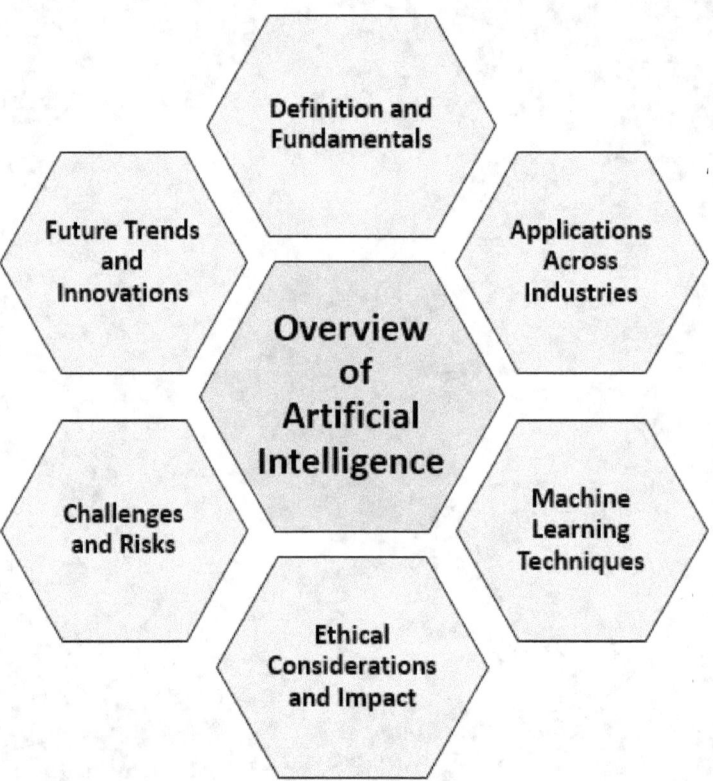

Beginning with the definition and fundamentals, it introduces core concepts such as machine learning, algorithms, and artificial neural networks. The exploration of applications across industries emphasizes industry-specific use cases, automation, and optimization. Machine learning techniques, including supervised and unsupervised learning, are highlighted for their pivotal role in AI applications. Ethical considerations, encompassing bias mitigation, transparency, and societal impact, underscore the responsible use of AI technologies. Identification of challenges and risks, from technical limitations to security concerns and job displacement, provides a realistic perspective. Lastly, the overview delves into future trends and innovations, exploring emerging technologies, research directions, and overarching AI trends. Together, these facets contribute to a holistic understanding of AI's current impact and its potential trajectory in diverse industries.

AI encompasses various approaches, including symbolic reasoning, machine learning, and neural networks. Symbolic AI relies on explicit programming and knowledge representation, while machine learning leverages statistical techniques to enable systems to learn from data. Neural networks, inspired by the human brain's structure, play a crucial role in deep learning—a subset of machine learning that involves complex, hierarchical data representations.

The applications of AI are vast and diverse, spanning industries such as healthcare, finance, manufacturing, and beyond. In healthcare, AI aids in medical diagnosis, personalized treatment plans, and drug discovery. Financial institutions leverage AI for fraud detection, risk assessment, and algorithmic trading. In manufacturing, AI contributes to process optimization, predictive maintenance, and quality control.

Natural Language Processing (NLP) is another critical aspect of AI, enabling machines to understand, interpret, and generate human language. This capability drives advancements in virtual assistants, chatbots, and language translation, enhancing human-computer interactions and communication.

Ethical considerations are integral to the development and deployment of AI. Issues such as bias in algorithms, transparency, accountability, and data privacy demand careful scrutiny. Striking a balance between technological innovation and ethical responsibility is essential to ensure AI benefits society while minimizing potential risks.

As AI continues to evolve, its impact on society, economies, and various sectors is poised to grow. The ongoing exploration of AI's potential, combined with ethical frameworks and responsible development practices, will shape its role in shaping the future of technology and human-machine interactions.

## 1.2 AI's Relevance to Civil Engineering

Artificial Intelligence (AI) holds significant relevance to the field of civil engineering, ushering in a new era of innovation and efficiency.

Its integration transforms traditional practices, offering solutions to complex challenges and augmenting the capabilities of civil engineers. The following points highlight AI's relevance to civil engineering:

1. **Optimized Project Planning and Design:** AI facilitates data-driven decision-making in project planning and design. Through predictive analytics and optimization algorithms, it helps engineers analyze historical data, identify patterns, and make informed choices, leading to more efficient project plans and designs.

2. **Enhanced Structural Analysis:** AI-driven tools excel in structural analysis, aiding engineers in assessing the integrity and safety of structures. Machine learning algorithms can process immense datasets to predict potential structural issues, contributing to more accurate risk assessments.

3. **Construction Project Management:** AI streamlines project management processes by offering predictive insights into resource allocation, scheduling, and risk management. It helps in identifying potential delays, optimizing workflows, and enhancing overall project efficiency.

4. **Smart Infrastructure and Asset Management:** AI contributes to the development of smart infrastructure by integrating with the Internet of Things (IoT). This collaboration enables real-time monitoring of assets, predictive maintenance, and the efficient management of infrastructure components, ensuring longevity and performance.

5. **Energy Efficiency and Sustainability:** AI plays a pivotal role in promoting sustainability within civil engineering. By analyzing data related to energy consumption, waste management, and environmental impact, AI assists in designing eco-friendly structures and implementing energy-efficient solutions.

6. **Robotics in Construction:** The use of robotics, a subset of AI, in construction operations enhances efficiency and safety. Autonomous construction vehicles, drones, and robotic arms contribute to tasks such as excavation, material handling, and site inspection, reducing manual labor and minimizing risks.

7. **Real-time Monitoring and Decision Support:** AI's ability to process and analyze data in real-time allows for continuous monitoring of construction projects. This enables quick identification of deviations from plans and facilitates timely decision-making to address issues as they arise.

8. **Quality Control and Assurance:** AI-powered systems contribute to quality control by automating inspection processes. Computer vision technologies can detect defects or deviations from specifications, ensuring that construction meets the required standards.

9. **Advanced Geotechnical and Environmental Analysis:** AI aids in geotechnical and environmental assessments by analyzing soil compositions, identifying potential hazards, and predicting environmental impacts. This information is crucial for informed decision-making in site selection and project execution.

10. **Human-Machine Collaboration:** The collaboration between human expertise and AI capabilities, often referred to as augmented intelligence, is changing the dynamics of civil engineering. Engineers can leverage AI tools for data analysis, leaving them more time for creative problem-solving and strategic decision-making.

In summary, AI's relevance to civil engineering lies in its capacity to enhance efficiency, accuracy, and sustainability across various aspects of the field. As technology continues to advance, the integration of AI is poised to play an increasingly pivotal role in shaping the future of civil engineering practices.

## 1.3 Historical Development of AI in civil engineering and construction

The historical development of Artificial Intelligence (AI) in civil engineering and construction reflects a gradual integration of advanced technologies to enhance efficiency, precision, and decision-making processes within the industry.

The timeline below outlines key milestones in the evolution of AI in this field:

1.  **1950s-1960s: Foundation of AI**

1. The inception of AI as a field of study occurred in the mid-20th century, laying the foundation for the exploration of intelligent machines capable of simulating human cognition.

2. Early AI research focused on symbolic reasoning, rule-based systems, and problem-solving algorithms, providing the groundwork for future applications in civil engineering.

2. **1970s-1980s: Expert Systems and Knowledge-Based Systems**

   - The development of expert systems, a form of AI that emulates human expertise in a specific domain, gained prominence.

   - Expert systems were applied to civil engineering for tasks such as project scheduling, design optimization, and risk assessment.

   - Knowledge-based systems emerged, utilizing databases of expert knowledge to assist engineers in decision-making processes.

3. **1990s-2000s: Integration of AI in Structural Analysis and Design**

   - AI technologies, including neural networks and fuzzy logic, were increasingly applied to structural analysis and design optimization.

   - Structural health monitoring systems using AI algorithms began to emerge, enabling real-time assessment of structural integrity.

   - Early applications of AI in construction management focused on scheduling, cost estimation, and risk analysis.

4. **2000s-2010s: Machine Learning and Data Analytics**

   - The adoption of machine learning techniques became more prevalent in civil engineering and construction management.

   - AI applications extended to predictive analytics, aiding in forecasting project timelines, resource requirements, and potential risks.

- Building Information Modeling (BIM) incorporated AI algorithms for enhanced collaboration and data-driven decision-making.

5. **2010s-Present: Advanced Applications and Robotics**

- The integration of AI with robotics became a significant trend in construction, with autonomous vehicles, drones, and robotic systems aiding in various construction tasks.

- AI-driven solutions expanded to include virtual and augmented reality for design visualization, simulation, and project communication.

- Advanced data analytics and AI algorithms are employed in real-time project monitoring, enabling adaptive decision-making during construction phases.

6. **Future Directions: AI and Smart Infrastructure**

- Ongoing research and development focus on integrating AI with the Internet of Things (IoT) for the creation of smart infrastructure.

- AI is expected to play a crucial role in developing sustainable and energy-efficient construction practices.

- Continued advancements in machine learning, natural language processing, and computer vision are anticipated to further revolutionize how civil engineering projects are conceptualized, executed, and maintained.

The historical trajectory of AI in civil engineering and construction management showcases a progression from foundational concepts to practical applications that are increasingly shaping the industry's future. As AI technologies continue to evolve, their integration is poised to bring about further innovations, efficiencies, and advancements in the construction and infrastructure development landscape.

## 1.4 Current State of AI

The current state of AI in construction management reflects a dynamic landscape marked by ongoing advancements and increasing adoption of artificial intelligence technologies within the industry.

Here are key aspects that characterize the current state of AI in construction management:

1. **Project Planning and Scheduling:**
   - AI is widely utilized for project planning and scheduling, offering advanced algorithms for optimizing timelines, resource allocation, and task sequencing.

- Predictive analytics assists in forecasting potential delays and facilitating proactive decision-making to keep projects on track.

2. **Risk Management and Predictive Analytics:**

- AI tools contribute significantly to risk management by analyzing historical project data and identifying potential risks.

- Predictive analytics models help construction managers anticipate challenges, enabling them to implement mitigation strategies and improve project outcomes.

3. **Construction Equipment and Robotics:**

- Robotics and autonomous construction equipment, often integrated with AI, enhance efficiency and safety on construction sites.

- Drones equipped with AI algorithms are utilized for surveying, mapping, and monitoring construction progress.

4. **Virtual Design and Construction (VDC):**

- AI plays a vital role in Virtual Design and Construction processes, aiding in 3D modeling, clash detection, and design optimization.

- BIM (Building Information Modeling) platforms incorporate AI algorithms for improved collaboration and data-driven decision-making.

5. **Supply Chain Management:**

- AI is applied to optimize supply chain management in construction by predicting material requirements, managing inventory, and ensuring timely deliveries.

- Smart logistics systems powered by AI contribute to more efficient transportation and distribution of construction materials.

6. **Quality Control and Safety:**

- Computer vision and AI algorithms are used for quality control, identifying defects or deviations from specifications during construction processes.

- AI-driven safety systems contribute to the monitoring of worksites, detecting potential hazards and ensuring compliance with safety protocols.

7. **Data Analytics for Decision Support:**

- Construction managers leverage AI-driven data analytics for informed decision-making.

- Real-time data analysis provides insights into project performance, enabling quick adjustments and enhancing overall project control.

8. **Human-Machine Collaboration:**

- Augmented intelligence, emphasizing collaboration between humans and AI, is increasingly prevalent.

- Construction professionals use AI tools for data analysis, leaving them more time for strategic decision-making and creative problem-solving.

9. **Documentation and Reporting:**

- AI technologies automate documentation processes, generating reports, and summarizing project data.

- Natural Language Processing (NLP) facilitates easier interaction with project data through voice commands and text-based queries.

10. **Adoption Challenges and Ethical Considerations:**

- Despite the advancements, challenges related to data privacy, security, and the need for skilled AI professionals persist.

- Ethical considerations, including bias in algorithms and the responsible use of AI, are important factors influencing AI adoption in construction management.

## 1.5 Future Trends and Challenges

The future trends and challenges in the application of AI in civil engineering and construction management:

**Future Trends:**

1. **AI-Driven Smart Infrastructure:**

   - The integration of AI with the Internet of Things (IoT) is expected to lead to the development of smart infrastructure with real-time monitoring and adaptive capabilities.

2. **Generative Design and AI-Assisted Planning:**

   - Generative design, coupled with AI algorithms, is likely to become more prevalent in creating optimized and innovative design solutions, influencing the early stages of project planning.

3. **AI for Sustainable Construction:**

   - AI technologies will play a pivotal role in advancing sustainable construction practices, optimizing energy usage, and reducing environmental impact through data-driven decision-making.

4. **Autonomous Construction Vehicles and Robotics:**

   - Continued advancements in robotics and AI will lead to increased use of autonomous construction vehicles and robotic systems for various tasks, reducing manual labor and enhancing efficiency.

5. **Enhanced Human-Machine Collaboration:**

   - Augmented intelligence will see further development, emphasizing the collaboration between AI systems and human professionals, enhancing decision-making and problem-solving capabilities.

6. **Advancements in Predictive Analytics:**

   - Predictive analytics models will become more sophisticated, offering construction managers better insights into project risks, costs, and timelines.

7. **Blockchain Integration for Data Security:**

   - The integration of blockchain technology may gain traction to enhance data security, transparency, and traceability in construction projects, addressing concerns related to data privacy.

8. **Edge Computing for Real-Time Analysis:**

   - Edge computing will be increasingly utilized for real-time data analysis on construction sites, reducing latency and improving the responsiveness of AI applications.

9. **Continued Growth of Digital Twins:**

- Digital twins, virtual replicas of physical assets or systems, will see expanded use in construction, allowing for better monitoring, simulation, and optimization throughout the project lifecycle.

10. **Customization of AI Solutions:**

- AI solutions will become more customizable, allowing construction firms to tailor applications to their specific needs, leading to greater adoption across various project types and scales.

**Challenges:**

1. **Data Quality and Availability:**

- The quality and availability of data remain a significant challenge. Obtaining accurate and reliable data for AI applications can be challenging, especially in large and complex construction projects.

2. **Integration with Existing Systems:**

- Integrating AI into existing workflows and systems may pose challenges, requiring careful planning and potentially causing disruptions during the implementation phase.

3. **Ethical Considerations and Bias:**

- Addressing ethical considerations, such as bias in algorithms, transparency, and accountability, will continue to be a crucial challenge in the responsible deployment of AI in construction.

4. **Skilled Workforce:**

- The need for a skilled workforce proficient in AI technologies is a challenge. The industry may face a shortage of professionals with expertise in both construction management and AI.

5. **Regulatory and Legal Frameworks:**

- Developing robust regulatory and legal frameworks to govern the use of AI in construction, addressing issues of liability and compliance, will be essential for widespread adoption.

6. **Initial Investment Costs:**

- The initial costs associated with implementing AI technologies can be a barrier for some construction firms. Demonstrating the long-term benefits and return on investment is crucial.

7. **Data Privacy and Security:**

- Protecting sensitive project data and ensuring data privacy will be an ongoing challenge, especially with the increasing amount of data being generated and processed.

8. **Interoperability Issues:**

- Ensuring interoperability between different AI systems and technologies used by various stakeholders in the construction process may pose challenges for seamless collaboration.

9. **Adaptation to Rapid Technological Changes:**

- The construction industry must adapt to the rapid pace of technological changes in AI, necessitating continuous learning and staying updated on emerging trends.

10. **Resistance to Change:**

- Resistance to adopting AI technologies, whether from industry professionals or organizational structures, can impede the smooth integration of AI in construction processes.

Addressing these challenges while capitalizing on emerging trends will be essential for the successful and responsible integration of AI in civil engineering and construction management in the future.

# 1.6 Real-world Examples / Use Cases

1. **Autodesk's Generative Design:**

   - Autodesk, a leading design software company, uses AI in its Generative Design tool to create thousands of design options based on specified constraints, helping engineers optimize structures for various factors like cost, material usage, and structural integrity.

2. **Caterpillar's AI-Enabled Construction Equipment:**

   - Caterpillar incorporates AI in construction equipment to enable autonomous operation, enhancing efficiency and safety on construction sites. AI algorithms assist in navigation, obstacle detection, and equipment coordination.

3. **Bentley Systems:**

   - Bentley Systems employs AI in its infrastructure design software to optimize project workflows, allowing civil engineers to leverage AI for design automation, simulation, and performance analysis.

   - Bechtel, a global engineering and construction company, historically integrated expert systems in the 1980s for decision support in project management, marking an early application of AI in the industry.

4. **Procore's Construction Management Software:**

   - Procore, a construction management software provider, integrates AI in its platform to streamline project management processes. AI features include predictive analytics for project risk assessment, schedule optimization, and budget forecasting.

5. **Predictive Maintenance by Komatsu:**

   - Komatsu, a construction equipment manufacturer, uses AI for predictive maintenance. AI algorithms analyze equipment data to predict potential failures, enabling proactive maintenance and minimizing downtime.

# 1.7 Chapter Summary: Key Points

1. Artificial Intelligence (AI) is transforming civil engineering, introducing innovative problem-solving approaches and reshaping traditional practices.

2. In recent years, AI integration in civil engineering has gained momentum due to project complexity, fostering proactive and data-driven solutions.

3. AI's analytical prowess allows rapid and accurate data processing, identifying patterns, trends, and potential risks overlooked by human observation

4. AI applications in civil engineering span project planning, design optimization, structural health monitoring, and real-time project control.

5. AI-driven technologies, including robotics and autonomous vehicles, enhance on-site operations, promoting safety, precision, and cost-effectiveness.

6. Ethical considerations, such as bias in algorithms and job displacement concerns, must be carefully addressed in AI adoption for responsible integration.

7. AI's collaboration with virtual and augmented reality adds a new dimension to project design and visualization, fostering better communication among stakeholders.

8. AI contributes significantly to sustainability in civil engineering, analyzing environmental data for eco-friendly infrastructure development.

9. Real-time monitoring powered by AI enables project managers to identify bottlenecks, risks, and deviations, facilitating timely interventions.

10. The evolving role of AI necessitates a shift in workforce skillsets towards augmented intelligence, emphasizing collaboration between human intuition and AI analytics.

11. Challenges in AI adoption include data security, mitigating algorithmic biases, and addressing concerns related to job displacement.

12. The integration of AI in civil engineering follows a systematic nine-stage process, from data collection to post-construction monitoring and evaluation.

13. The future of AI in civil engineering holds exciting prospects, including smart infrastructure, generative design, sustainable construction, and enhanced human-machine collaboration.

14. Ongoing challenges in AI adoption encompass data quality, integration with existing systems, ethical considerations, and the need for a skilled workforce.

15. Successful integration requires addressing regulatory frameworks, demonstrating long-term benefits, ensuring data privacy, and overcoming resistance to change in the construction industry.

# 1.8 Concept Check: Q&A Sessions

1. How does artificial intelligence (AI) contribute to optimized project planning and design in civil engineering?

2. What role does AI play in enhancing structural analysis and ensuring the safety of civil engineering structures?

3. How does AI streamline project management processes, including resource allocation, scheduling, and risk management, in construction projects?

4. In what ways does AI contribute to the development of smart infrastructure and efficient asset management within civil engineering?

5. How does AI promote energy efficiency and sustainability in the field of civil engineering, particularly in designing eco-friendly structures?

6. What are the applications of robotics, a subset of AI, in construction operations, and how do they enhance efficiency and safety on construction sites?

7. How does AI facilitate real-time monitoring and decision support in construction projects, enabling quick identification of deviations from plans?

8. What role does AI-powered systems play in quality control during construction processes, and how do they ensure that construction meets required standards?

9. How does AI aid in geotechnical and environmental analysis in civil engineering, contributing to informed decision-making in site selection and project execution?

10. How is the concept of augmented intelligence, emphasizing collaboration between humans and AI, changing the dynamics of civil engineering practices?

# 2.0 Fundamentals of Civil Engineering

Civil engineering is a branch of engineering that encompasses the design, construction, and maintenance of infrastructure, including buildings, bridges, roads, dams, and water supply systems.

It plays a pivotal role in shaping the physical environment and enhancing the quality of life. The profession involves a deep understanding of mathematics, physics, and materials science to devise solutions for complex engineering challenges.

One fundamental aspect of civil engineering is structural engineering, focusing on the design and analysis of structures to ensure they withstand various loads and environmental conditions. Geotechnical engineering deals with the behavior of soil and rock, essential for foundation design and slope stability. Transportation engineering involves planning and designing efficient and safe transportation systems, encompassing roads, railways, and airports.

Hydraulic engineering is crucial for managing water resources, addressing issues such as flood control, irrigation, and water supply. Environmental engineering emphasizes sustainable practices, addressing pollution control and waste management. Surveying and mapping play an integral role in collecting and interpreting spatial data for accurate project planning.

Civil engineering is a multifaceted discipline comprised of following several key areas of specialization.

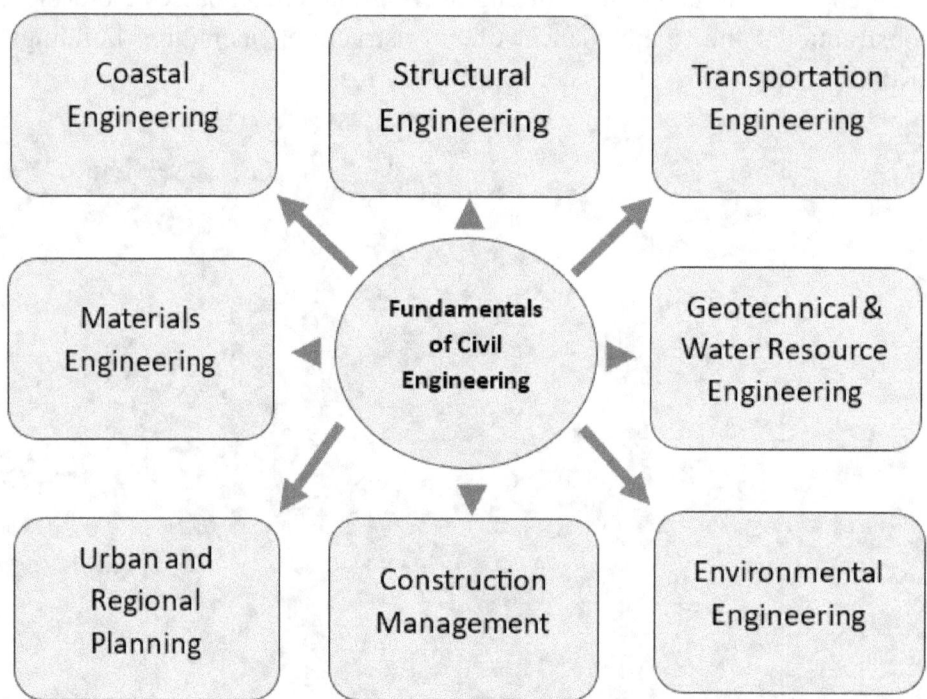

Structural engineering focuses on designing and analyzing structures for durability in various conditions, including buildings, bridges, and dams. Transportation engineering involves planning and managing efficient transportation systems, encompassing roads, railways, airports, and public transit. Geotechnical engineering studies soil and rock properties for construction suitability, applicable to foundations, tunnels, and retaining walls. Water resources engineering manages water distribution and control, covering dams, water treatment plants, and drainage systems. Environmental engineering addresses environmental aspects of projects, such as pollution control and waste management. Construction management involves planning and executing projects within budget and timeline constraints. Urban and regional planning optimizes land use for population growth and infrastructure development. Materials engineering focuses on construction material properties and applications. Coastal engineering deals with coastal area design, including erosion prevention and infrastructure resilience. Interdisciplinary collaboration is common, and evolving areas like smart infrastructure and sustainable development reflect the dynamic nature of civil engineering in shaping the built environment to meet society's evolving needs.

Construction management involves overseeing the entire construction process, coordinating resources, and ensuring projects are completed on time and within budget. Materials engineering focuses on selecting and testing materials for construction, considering factors like durability and cost-effectiveness. Coastal engineering deals with the challenges posed by coastal environments, including erosion control and the design of coastal structures.

Moreover, civil engineers must adhere to ethical standards, considering the safety and well-being of the public. Effective communication and collaboration with architects, contractors, and other stakeholders are essential for successful project execution. In essence, civil engineering serves as a cornerstone for societal development, addressing the intricate balance between technological advancements, environmental sustainability, and public welfare.

Further, fundamentals of civil engineering encompass a broad range of principles, theories, and practices that form the foundation of the discipline. These fundamentals provide the basis for designing, planning, constructing, and maintaining infrastructure projects. Here are key aspects of the fundamentals of civil engineering:

1. **Mathematics and Physics:**

   - Civil engineering relies heavily on mathematical and physical principles. Concepts such as calculus, algebra, geometry, and physics are essential for analyzing and solving engineering problems.

2. **Statics and Dynamics:**

   - Understanding statics (equilibrium of stationary objects) and dynamics (motion and acceleration) is crucial for designing structures that can withstand various loads and environmental conditions.

3. **Mechanics of Materials:**

   - Mechanics of materials explores how materials respond to various forces. This includes the study of stresses, strains, and material properties, essential for designing structures and components.

4. **Fluid Mechanics:**
   - Knowledge of fluid mechanics is vital for designing hydraulic structures, pipelines, and drainage systems. It involves the study of fluid behavior under different conditions.

5. **Geotechnical Engineering:**
   - Geotechnical engineering focuses on the behavior of soils and rocks. It includes understanding soil properties, foundation design, and slope stability analysis.

6. **Structural Analysis and Design:**
   - Structural analysis involves studying how structures react to applied forces, while structural design involves creating safe and efficient structures, considering material properties and structural integrity.

7. **Transportation Engineering:**
   - Transportation engineering deals with the planning, design, and maintenance of transportation systems, including roads, highways, railways, airports, and public transit.

8. **Environmental Engineering:**
   - Environmental engineering addresses the impact of engineering projects on the environment. It includes water and air quality management, waste treatment, and sustainable practices.

9. **Surveying and Geomatics:**
   - Surveying involves measuring and mapping the Earth's surface, providing essential data for designing and constructing projects. Geomatics incorporates GIS (Geographic Information Systems) and remote sensing technologies.

10. **Construction Management:**
    - Construction management involves planning, scheduling, and overseeing the construction process. It includes project management principles, cost estimation, and quality control.

11. **Materials Science:**

- Understanding the properties and behavior of construction materials, such as concrete, steel, wood, and composites, is crucial for selecting appropriate materials in design and construction.

### 12. Hydraulics and Water Resources Engineering:

- Hydraulics deals with the flow of fluids, including water. Water resources engineering involves managing water-related projects, such as dams, irrigation systems, and water supply networks.

### 13. Engineering Ethics and Professional Practice:

- Civil engineers adhere to a code of ethics that guides professional conduct. This includes considerations of public safety, environmental impact, and the well-being of communities.

### 14. Risk Assessment and Management:

- Evaluating and managing risks associated with engineering projects is fundamental. This involves identifying potential hazards, assessing their likelihood and consequences, and implementing risk mitigation strategies.

### 15. Computer-Aided Design (CAD) and Modeling:

- Proficiency in CAD software and modeling tools is essential for creating detailed designs, simulations, and visualizations of engineering projects.

### 16. Sustainability Principles:

- Considering environmental, economic, and social sustainability is increasingly important in civil engineering. Sustainable practices aim to minimize the environmental impact of projects and ensure long-term viability.

These fundamentals collectively provide the knowledge and skills required for civil engineers to analyze, design, and implement a wide array of infrastructure projects, contributing to the improvement and sustainability of societies.

## 2.1 Core Principles in Civil Engineering

The core principles in civil engineering form the foundational concepts that guide the profession.

These principles are fundamental to understanding and addressing the challenges involved in planning, designing, constructing, and maintaining infrastructure projects. Here are the core principles in civil engineering:

1. **Safety:**

    - The paramount principle in civil engineering is ensuring the safety of structures, systems, and the public. Engineers must design and construct projects with a focus on minimizing risks and preventing failures.

2. **Sustainability:**

- Sustainable practices aim to meet the needs of the present without compromising the ability of future generations to meet their own needs. Civil engineers consider environmental, economic, and social sustainability in their designs and decisions.

3. **Functionality and Utility:**

- Civil engineering projects should fulfill their intended purpose and provide utility to society. This principle emphasizes the importance of meeting the functional requirements of infrastructure projects.

4. **Economic Viability:**

- Engineers consider the economic feasibility of projects, taking into account factors such as cost-effectiveness, life-cycle costs, and return on investment. Balancing economic considerations ensures efficient resource allocation.

5. **Durability:**

- Structures and infrastructure should be designed for long-term durability and resilience. Considering material properties, environmental conditions, and maintenance requirements contributes to the longevity of projects.

6. **Ethical Practices:**

- Engineers adhere to a strict code of ethics, prioritizing integrity, honesty, and accountability. Ethical considerations guide decision-making and professional conduct in the best interest of society.

7. **Environmental Impact:**

- Civil engineers assess and mitigate the environmental impact of projects. Minimizing negative effects on ecosystems, air, water, and soil quality is essential for responsible engineering practice.

8. **Public Health and Welfare:**

- Engineers prioritize the well-being of the public. Projects should enhance public safety, health, and overall welfare, reflecting a commitment to serving the interests of communities.

9. **Innovation and Creativity:**

- Encouraging innovation and creative problem-solving is a core principle. Engineers seek new and improved solutions to address challenges, fostering advancements in the field.

10. **Adaptability and Resilience:**

- Designs should consider adaptability to changing conditions and potential hazards. Building resilience into infrastructure ensures it can withstand unforeseen events and challenges.

11. **Professional Development and Continuing Education:**

- Engineers commit to ongoing professional development and continuing education to stay abreast of technological advancements, industry best practices, and emerging trends in civil engineering.

12. **Regulatory Compliance:**

- Engineers adhere to local, national, and international regulations and codes. Ensuring compliance with standards and regulations is essential for the safety and legality of projects.

13. **Interdisciplinary Collaboration:**

- Civil engineers often collaborate with professionals from other disciplines, recognizing the interconnectedness of various fields. Interdisciplinary collaboration enhances the comprehensive understanding and execution of projects.

14. **Community Engagement:**

- Engaging with the community and stakeholders is vital. Understanding the needs, concerns, and aspirations of communities helps shape projects that align with societal values and expectations.

15. **Quality Assurance and Quality Control:**

- Implementing rigorous quality assurance and quality control processes is crucial. Engineers aim to deliver projects that meet or exceed specified standards and requirements.

These core principles serve as a guide for ethical, responsible, and effective civil engineering practice, ensuring that engineers contribute positively to the well-being of society and the sustainable development of infrastructure.

## 2.2 Traditional Construction Practices

Traditional construction practices refer to established methods and techniques that have been used for generations in the construction of buildings and infrastructure.

These practices are often based on local materials, craftsmanship, and cultural influences. While modern construction methods and technologies have evolved, traditional practices continue to be relevant, especially in regions where heritage, cultural preservation, and availability of traditional materials play a significant role. Here are some aspects of traditional construction practices:

1. **Masonry Construction:**

- Traditional masonry involves the use of locally sourced materials such as stone, brick, or adobe, often combined with mortar. Masonry structures, including walls and arches, are built by skilled craftsmen using time-tested techniques.

2. **Timber Framing:**

   - Timber framing is a traditional construction method where wooden frames are assembled using joints and pegs rather than nails. This method is often associated with traditional houses, barns, and structures.

3. **Thatch Roofing:**

   - Thatch roofing, made from dry vegetation such as straw or reeds, is a traditional roofing method used in many cultures. Thatch provides natural insulation and has a distinctive appearance.

4. **Adobe Construction:**

   - Adobe is a traditional building material made from sun-dried clay, straw, and water. Adobe bricks are formed and dried in the sun before being used to construct walls, often in regions with a dry climate.

5. **Cob Construction:**

   - Cob is a mixture of clay, sand, and straw that is used to build walls. Cob construction involves shaping the material into monolithic structures, and it has been historically used for homes and small structures.

6. **Rammed Earth Construction:**

   - Rammed earth construction involves compressing a mixture of earth, chalk, lime, or gravel within a frame to create sturdy walls. This method is known for its thermal mass properties.

7. **Bamboo Construction:**

   - In regions where bamboo is abundant, it has been a traditional construction material for various structures. Bamboo's strength and flexibility make it suitable for scaffolding, walls, and even structural components.

8. **Dry Stone Walling:**

   - Dry stone walling is a technique where stones are stacked without mortar. This method is often used for retaining walls, boundary walls, and agricultural structures.

9. **Wattle and Daub:**

   - Wattle and daub construction involves weaving thin branches or sticks (wattle) and covering them with a mixture of mud, clay, or other materials (daub). This method is often used for walls and partitions.

10. **Post and Beam Construction:**

    - Post and beam construction uses vertical posts to support horizontal beams, creating a structural framework. This method is often associated with traditional barns and large open interior spaces.

11. **Lime Plastering:**

    - Lime plaster, made from slaked lime, sand, and water, has been traditionally used as a finish for walls. It provides a breathable surface and has antimicrobial properties.

12. **Stone Arch Construction:**

    - Stone arches are a traditional method of constructing load-bearing structures, often seen in bridges, aqueducts, and historic buildings.

13. **Terracotta Tile Roofing:**

    - Terracotta tiles, made from baked clay, have been used for centuries as roofing materials in various regions. These tiles provide durability and effective rainwater runoff.

14. **Cantilevered Construction:**

    - Cantilevered structures, where a beam or slab is supported on only one end, have been used in traditional architecture for creating overhangs and balconies.

15. **Vaulted Construction:**

- Vaulted construction involves creating arched structures or ceilings, often seen in traditional churches, cathedrals, and historic buildings.

While these traditional construction practices have historical and cultural significance, it's important to note that they coexist with and are sometimes integrated into modern construction methods, combining the benefits of tradition with contemporary engineering practices. Additionally, the preservation of traditional techniques contributes to the conservation of cultural heritage and sustainable building practices.

## 2.3 Emerging Technologies

Emerging technologies are continually shaping the landscape of civil engineering and construction engineering and management, driving innovation, efficiency, and sustainability.

Here are some of the notable emerging technologies in these fields:

1. **Building Information Modeling (BIM):**

   - BIM involves creating digital representations of the physical and functional characteristics of a building or infrastructure. It facilitates collaboration, improves project visualization, and enhances decision-making throughout the project lifecycle.

2. **Artificial Intelligence (AI) and Machine Learning (ML):**

- AI and ML are increasingly applied in civil engineering for tasks such as predictive analytics, project scheduling, design optimization, and risk management. These technologies enable data-driven decision-making and can enhance project efficiency.

3. **Augmented Reality (AR) and Virtual Reality (VR):**

- AR and VR technologies are used for immersive project visualization, design reviews, and construction planning. They offer enhanced collaboration, training, and simulation experiences for construction professionals.

4. **Drones and Unmanned Aerial Vehicles (UAVs):**

- Drones provide aerial surveys, site inspections, and data collection, offering a cost-effective and efficient way to monitor construction sites, assess progress, and gather geospatial information.

5. **3D Printing (Additive Manufacturing):**

- 3D printing is being explored for construction purposes, allowing the on-site fabrication of building components. This technology has the potential to reduce material waste and enhance construction speed.

6. **Robotics in Construction:**

- Robotics is utilized for tasks such as bricklaying, concrete pouring, and site inspection. Autonomous construction vehicles and robotic systems contribute to increased efficiency and safety on construction sites.

7. **Smart Sensors and IoT (Internet of Things):**

- Smart sensors and IoT devices are used for real-time monitoring of structural health, environmental conditions, and equipment performance. This data helps in predictive maintenance and ensures the longevity of infrastructure.

8. **5G Technology:**

- The rollout of 5G networks enables faster and more reliable communication on construction sites. It supports real-time data transfer, remote monitoring, and the connectivity of a large number of devices.

9. **Self-Healing Materials:**

   - Self-healing materials are designed to repair damage autonomously, extending the lifespan of structures and reducing maintenance costs. This technology is particularly valuable in enhancing the durability of infrastructure.

10. **Smart Grids and Sustainable Energy Solutions:**

    - Integration of smart grids and sustainable energy solutions in construction projects contributes to energy efficiency. Technologies like solar panels, energy storage systems, and smart building controls are increasingly common.

11. **Blockchain Technology:**

    - Blockchain is explored for its potential in improving transparency, security, and efficiency in construction project management. It can be used for secure record-keeping, supply chain management, and financial transactions.

12. **Generative Design:**

    - Generative design uses algorithms to explore a multitude of design options based on specified criteria. This technology assists engineers in creating optimized and innovative designs.

13. **Hybrid Reality:**

    - Hybrid reality combines physical and digital elements, allowing real-world and digital models to interact. This integration enhances collaboration, communication, and decision-making in construction projects.

14. **Digital Twins:**

    - Digital twins are virtual replicas of physical objects or systems. In civil engineering, digital twins are used for real-time monitoring, simulation, and analysis of infrastructure, facilitating better decision-making and maintenance.

15. **Nanotechnology in Construction Materials:**

- Nanotechnology is applied to enhance the properties of construction materials, such as improving strength, durability, and resistance to environmental factors.

The integration of these emerging technologies represents a transformative shift in the way civil engineering and construction projects are conceived, executed, and managed. Continued research and development in these areas are expected to further revolutionize the industry, making construction processes more efficient, sustainable, and resilient.

## 2.4 Integration of AI in Civil Engineering

The integration of Artificial Intelligence (AI) in civil engineering, construction engineering, and management is transforming the industry by enhancing efficiency, decision-making, and overall project outcomes.

Here are key aspects of how AI is integrated into these domains:

1. **Project Planning and Scheduling:**

   - **AI Algorithms for Optimization:** AI is used to optimize project schedules, resource allocation, and task sequencing. Machine learning algorithms can analyze historical project data to identify patterns and predict the most efficient project timelines.

2. **Design and Modeling:**

- **Generative Design:** AI-driven generative design tools help in creating and exploring numerous design options based on specified parameters, allowing engineers to identify optimal solutions.

- **Computer-Aided Design (CAD):** AI enhances CAD systems, automating certain design tasks, improving accuracy, and speeding up the design process.

3. **Structural Analysis and Design:**

- **AI for Structural Health Monitoring:** AI algorithms analyze real-time data from sensors to assess the health and integrity of structures, identifying potential issues and recommending maintenance or repairs.

- **Optimization of Structural Design:** AI is employed to optimize structural designs by considering various factors such as material usage, cost, and environmental impact.

4. **Construction Project Management:**

- **Predictive Analytics:** AI facilitates predictive analytics for project management, helping in forecasting potential risks, delays, and resource requirements.

- **Automated Scheduling:** AI algorithms automate project scheduling by considering various constraints, dependencies, and resource availability.

5. **Cost Estimation and Budgeting:**

- **AI-Based Cost Estimation:** AI algorithms analyze historical cost data, project specifications, and other factors to provide accurate cost estimates. This aids in budgeting and financial planning.

- **Expense Prediction:** AI helps predict potential cost overruns or deviations from the budget by analyzing project data in real-time.

6. **Supply Chain and Logistics:**

- **AI in Procurement:** AI is applied to optimize procurement processes, predict material requirements, and streamline the supply chain, ensuring timely deliveries.

- **Inventory Management:** AI helps in managing construction site inventories efficiently, reducing waste and ensuring the availability of materials as needed.

7. **Risk Management:**

- **Risk Prediction and Mitigation:** AI algorithms analyze historical project data and external factors to predict potential risks. This proactive approach allows for better risk mitigation strategies.

8. **Quality Control:**

- **Computer Vision for Inspection:** AI-powered computer vision systems are used for automated inspection of construction components, identifying defects or deviations from quality standards.

- **Data Analysis for Quality Trends:** AI analyzes data related to construction quality to identify trends, patterns, and areas for improvement.

9. **Human Resource Management:**

- **Skillset Analysis:** AI assists in analyzing the skillsets of the workforce, helping project managers make informed decisions about resource allocation and training needs.

- **Collaboration Tools:** AI-driven collaboration tools enhance communication and coordination among project team members.

10. **Real-time Monitoring and Reporting:**

- **IoT Integration:** AI is integrated with the Internet of Things (IoT) for real-time monitoring of construction sites, equipment, and environmental conditions.

- **Automated Reporting:** AI automates the generation of project reports, providing stakeholders with timely and accurate updates on project progress.

**11. Natural Language Processing (NLP):**

- **Voice-Activated Interfaces:** NLP enables voice-activated interfaces for project management software, facilitating hands-free interaction and improving accessibility on construction sites.

**12. Energy Efficiency and Sustainability:**

- **Energy Consumption Analysis:** AI analyzes data to optimize energy consumption in buildings and infrastructure, contributing to sustainable and energy-efficient designs.

- **Carbon Footprint Reduction:** AI is used to assess and minimize the environmental impact of construction projects, considering factors such as material choices and construction methods.

The integration of AI in civil engineering, construction engineering, and management is an ongoing process, with advancements continually expanding the capabilities of these technologies. As AI continues to evolve, its role in these fields is expected to further enhance productivity, reduce costs, and contribute to the development of sustainable and resilient infrastructure.

## 2.5 Benefits and Limitations of AI

Artificial intelligence provides various benefits in civil engineering and constructions along with few limitations.

**Benefits:**

1. **Improved Efficiency:**

   - AI automates routine tasks, speeding up processes such as project scheduling, resource allocation, and design optimization. This leads to increased overall efficiency in project delivery.

2. **Data-Driven Decision-Making:**

- AI processes vast amounts of data to provide insights and predictions, enabling more informed decision-making. This is particularly valuable in risk management, project planning, and resource optimization.

3. **Enhanced Safety:**

   - AI contributes to improved safety on construction sites through the use of robotics and autonomous vehicles. Drones and sensors equipped with AI can monitor sites for potential hazards, enhancing overall safety protocols.

4. **Optimized Design and Planning:**

   - Generative design powered by AI enables the exploration of numerous design options based on specified criteria, leading to more innovative and optimized solutions in the planning phase.

5. **Predictive Analytics for Risk Management:**

   - AI algorithms predict potential risks and delays by analyzing historical project data. This proactive approach allows for better risk management and mitigation strategies.

6. **Real-Time Monitoring:**

   - AI, in conjunction with IoT, facilitates real-time monitoring of construction sites, equipment, and environmental conditions. This ensures prompt identification of issues and enables timely interventions.

7. **Cost Savings:**

   - AI's ability to optimize resource allocation, reduce delays, and improve project efficiency contributes to cost savings. Accurate cost estimation and predictive analytics help in minimizing budget overruns.

8. **Quality Control and Inspection:**

   - AI-powered computer vision systems enhance quality control by automating the inspection process. This ensures that construction components meet specified standards, reducing the likelihood of defects.

9. **Smart Infrastructure and Sustainability:**

   - AI, integrated with the Internet of Things (IoT), contributes to the development of smart infrastructure. This includes real-time monitoring, predictive maintenance, and energy-efficient solutions, promoting sustainability.

10. **Human-Machine Collaboration:**

   - AI augments human capabilities by automating routine tasks, allowing professionals to focus on creative problem-solving and strategic decision-making. This collaboration enhances overall project outcomes.

## Limitations:

1. **Data Quality and Availability:**

   - The effectiveness of AI relies heavily on the quality and availability of data. Incomplete or inaccurate data can lead to suboptimal results and decisions.

2. **Initial Implementation Costs:**

   - The upfront costs associated with implementing AI technologies, including training and infrastructure, can be a barrier for some construction firms. Demonstrating the long-term benefits is crucial.

3. **Need for Skilled Workforce:**

   - Successfully implementing and managing AI systems requires a skilled workforce with expertise in both construction management and AI technologies. The industry may face a shortage of such professionals.

4. **Interoperability Challenges:**

   - Integrating AI with existing systems and ensuring interoperability with various software and platforms can be complex. This may require additional investments in technology integration.

5. **Ethical Considerations and Bias:**

- AI algorithms can inadvertently perpetuate biases present in historical data. Ensuring fairness and addressing ethical considerations in AI decision-making remains a challenge.

6. **Data Privacy and Security Concerns:**

- As AI systems handle sensitive project and personnel data, ensuring robust data privacy and security measures is essential to prevent unauthorized access and potential breaches.

7. **Regulatory and Legal Challenges:**

- The evolving nature of AI may pose challenges in terms of regulatory compliance and legal frameworks. Addressing issues related to liability and accountability is an ongoing consideration.

8. **Resistance to Adoption:**

- There may be resistance to adopting AI technologies, whether from industry professionals or organizational structures. Overcoming this resistance requires effective change management strategies.

9. **Lack of Standardization:**

- The absence of standardized practices and protocols for AI applications in construction may lead to variations in implementation and interoperability challenges.

10. **Complexity of AI Algorithms:**

- Understanding and managing complex AI algorithms may be challenging for some professionals. This complexity can hinder widespread adoption, particularly in smaller construction firms.

While the benefits of AI in civil engineering and construction management are substantial, addressing the associated limitations is crucial for successful implementation and maximizing the positive impact of these technologies. Ongoing research, collaboration, and advancements in AI ethics and standards are essential for the responsible integration of AI in the construction industry.

## 2.6 Real-world Examples / Use Cases

1. **Bricklaying Robot (Company: Fastbrick Robotics):**

   - Fastbrick Robotics has developed an AI-powered bricklaying robot named Hadrian X, capable of autonomously building brick structures. This technology enhances construction speed and precision.

2. **Predictive Analytics for Construction Delays (Company: Oracle Construction and Engineering):**

   - Oracle Construction and Engineering utilize AI-powered predictive analytics to forecast potential delays in construction projects. This technology helps project managers proactively address issues and optimize timelines.

3. **AI-Enhanced Construction Site Monitoring (Company: Skycatch):**

   - Skycatch combines drones and AI for real-time construction site monitoring. The system captures high-resolution images, and AI algorithms analyze the data to track progress, identify potential issues, and enhance project management.

4. **3D Printing of Buildings (Company: ICON):**

   - ICON utilizes 3D printing technology for construction, producing houses and structures using robotics. This emerging technology offers rapid and cost-effective construction solutions.

5. **Augmented Reality in Construction (Company: Trimble):**

   - Trimble's SiteVision uses augmented reality (AR) for construction site visualization. This technology overlays digital information onto the physical environment, aiding in design and project planning.

6. **Self-Healing Concrete (Company: Basilisk):**

   - Basilisk has developed self-healing concrete that uses bacteria to repair cracks autonomously. This emerging technology improves the durability and lifespan of concrete structures.

## 2.7 Chapter Summary: Key Points

1. Civil engineering encompasses the design, construction, and maintenance of infrastructure, addressing diverse areas like structural engineering, transportation, geotechnics, and environmental considerations.

2. Construction management involves coordinating resources to ensure projects are completed within budget and on time, emphasizing planning and execution efficiency.

3. Fundamental principles of civil engineering include a strong foundation in mathematics, physics, and key areas such as fluid mechanics, geotechnical engineering, and surveying.

4. Core principles in civil engineering prioritize safety, sustainability, functionality, and ethical practices, guiding professionals in responsible and effective project execution.

5. Traditional construction practices, rooted in local materials and cultural influences, coexist with modern methods, contributing to cultural heritage preservation.

6. Emerging technologies in civil engineering include BIM, AI, drones, 3D printing, and smart sensors, driving innovation, efficiency, and sustainability.

7. AI integration in civil engineering enhances project planning, design, structural analysis, and construction management through optimization algorithms and predictive analytics.

8. AI-driven technologies, such as generative design and computer vision, contribute to optimized design, real-time monitoring, and automated quality control in construction projects.

9. Benefits of AI in civil engineering include improved efficiency, data-driven decision-making, enhanced safety, optimized design, and real-time monitoring for cost savings and sustainability.

10. Limitations of AI in civil engineering involve challenges like data quality, initial implementation costs, the need for a skilled workforce, ethical considerations, and regulatory and legal complexities.

11. Interoperability challenges and resistance to adoption pose obstacles to widespread AI integration in the construction industry.

12. Addressing ethical concerns related to bias, data privacy, and security is crucial for responsible AI implementation in civil engineering.

13. Lack of standardization and the complexity of AI algorithms may hinder widespread adoption, particularly in smaller construction firms.

14. Ongoing research, collaboration, and advancements in AI ethics and standards are essential for the responsible integration of AI in the construction industry.

15. Balancing the benefits and limitations of AI ensures successful implementation and maximizes the positive impact of these technologies in civil engineering and construction management.

## 2.8 Concept Check: Q&A Sessions

1. What is the role of structural engineering in civil engineering, and why is it crucial for infrastructure projects?

2. Explain the significance of geotechnical engineering in the construction process, and how does it contribute to foundation design?

3. How does environmental engineering play a role in civil engineering projects, and what aspects does it address for sustainable practices?

4. In what ways do traditional construction practices, such as masonry and timber framing, coexist with modern construction methods, and why are they still relevant today?

5. How do emerging technologies, like Building Information Modeling (BIM) and Artificial Intelligence (AI), influence the landscape of civil engineering, and what benefits do they offer to the industry?

6. What are the key aspects of AI integration in civil engineering, and how does it transform project planning, design, and construction project management?

7. Examine the core principles in civil engineering, emphasizing safety, sustainability, and functionality. How do these principles guide the profession and project outcomes?

8. Discuss the benefits and limitations of AI in civil engineering. How does AI contribute to improved efficiency, safety, and decision-making, and what challenges does it present to the industry?

9. How does AI impact risk management in construction projects, and in what ways can AI algorithms predict and mitigate potential risks?

10. Explain the significance of sustainability principles in civil engineering and how AI is integrated into projects to promote energy efficiency and minimize environmental impact.

# 3. Basics of Artificial Intelligence

Artificial Intelligence (AI) refers to the development of computer systems capable of performing tasks that typically require human intelligence.

At its core, AI involves the creation of algorithms and models that enable machines to learn from data, recognize patterns, and make decisions autonomously. The primary objective of AI is to simulate human cognitive functions, such as problem-solving, learning, perception, and language understanding, within machines. Artificial Intelligence (AI) is a multidisciplinary field that involves the development of intelligent agents capable of performing tasks that typically require human intelligence.

The field of Artificial Intelligence (AI) is comprised of following eight fundamental components, forming a comprehensive framework for understanding its intricacies.

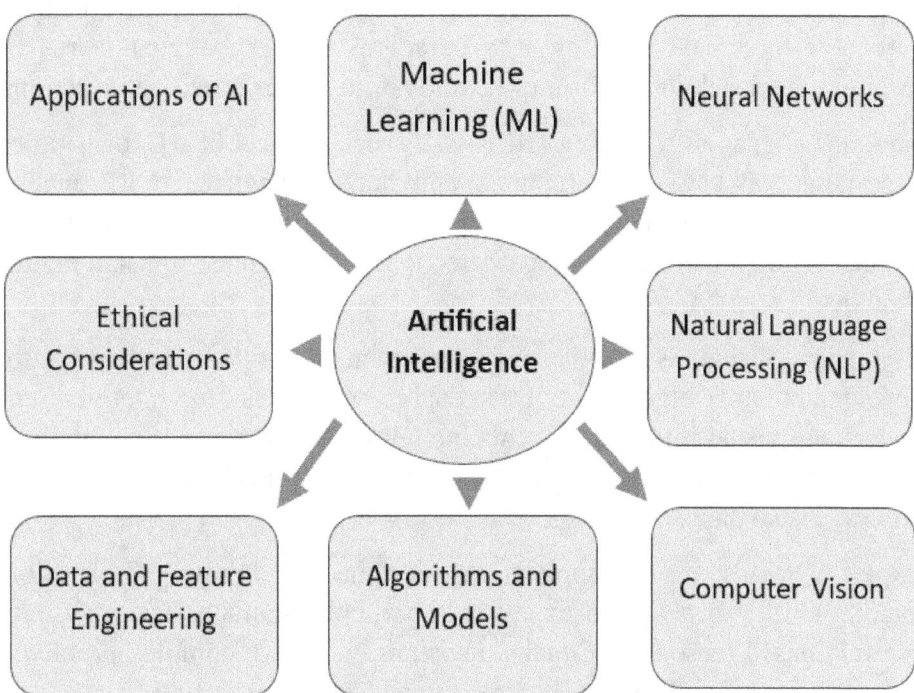

Beginning with an overarching definition and historical context of AI, the structure includes Machine Learning (ML), which encompasses supervised, unsupervised, and reinforcement learning. Neural Networks, covering basic principles and deep learning, represent a crucial aspect, alongside Natural Language Processing (NLP) and Computer Vision, which respectively delve into language comprehension and visual data processing applications. The inclusion of Algorithms and Models highlights decision trees, support vector machines, and deep neural networks. Data and Feature Engineering emphasize the pivotal role of data, encompassing its importance and techniques for feature selection and transformation. Ethical considerations such as bias, transparency, and responsible AI practices are integral components, and the mind map concludes with a focus on diverse applications of AI, spanning industry-specific contexts like healthcare and finance, as well as everyday applications such as virtual assistants and recommendation systems. This comprehensive structure facilitates a holistic understanding of AI's key elements and their interconnections.

Machine Learning is a foundational concept in AI, representing a subset where algorithms are designed to improve their performance over time through experience. In supervised learning, models are trained on labeled datasets, while unsupervised learning involves learning from unlabeled data. Reinforcement learning is a paradigm where an agent learns by receiving feedback in the form of rewards or penalties based on its actions.

Natural Language Processing (NLP) is a crucial aspect of AI, focusing on the interaction between computers and human language. NLP enables machines to understand, interpret, and generate human language, facilitating applications like speech recognition, language translation, and chatbots.

Computer Vision is another key component of AI, involving the development of algorithms that enable machines to interpret and understand visual information. This includes tasks like image recognition, object detection, and facial recognition, allowing machines to "see" and process visual data.

Expert Systems are AI applications designed to mimic the decision-making ability of a human expert in a particular domain. These systems use rule-based reasoning to make decisions and solve complex problems, contributing to fields like healthcare, finance, and engineering.

The Turing Test, proposed by Alan Turing, is a benchmark for assessing a machine's ability to exhibit human-like intelligence. If a machine can engage in conversation in a way indistinguishable from a human, it is considered to have passed the Turing Test.

AI applications are diverse, spanning industries such as healthcare, finance, education, and entertainment. In healthcare, AI aids in diagnostics, drug discovery, and personalized medicine. In finance, it contributes to fraud detection, algorithmic trading, and risk management.

Ethical considerations in AI are becoming increasingly important. Issues related to bias in algorithms, transparency, accountability, and the potential societal impact of AI technologies require careful attention to ensure responsible and fair deployment.

As AI technology continues to evolve, researchers are exploring advanced areas like Quantum Computing, which has the potential to significantly enhance the computational power available for AI applications.

The basics of Artificial Intelligence encompass concepts like Machine Learning, Natural Language Processing, Computer Vision, and Expert Systems. AI applications have widespread implications across industries, and ethical considerations are crucial in the responsible development and deployment of AI technologies. The field is dynamic, with ongoing advancements pushing the boundaries of what machines can achieve in emulating human intelligence.

Here are the key points on Basics of Artificial Intelligence:

### 1. Definition of AI:

Artificial Intelligence (AI): AI refers to the simulation of human intelligence in machines that are programmed to think and learn like humans. It encompasses a wide range of technologies and techniques.

### 2. Types of AI:

- **Narrow AI (Weak AI):** Designed and trained for a specific task, such as speech recognition or image classification.

- **General AI (Strong AI):** Possesses the ability to understand, learn, and apply knowledge across diverse tasks, similar to human intelligence (currently theoretical).

### 3. Machine Learning (ML):

A subset of AI that focuses on the development of algorithms and statistical models that enable computers to improve their performance on a task through learning from data.

### 4. Deep Learning:

A subfield of machine learning inspired by the structure and function of the human brain's neural networks. Deep learning involves neural networks with many layers (deep neural networks) to learn and make decisions.

### 5. Supervised Learning:

A type of machine learning where the algorithm is trained on a labeled dataset, meaning it learns from input-output pairs. It aims to make predictions or classifications.

### 6. Unsupervised Learning:

Involves training the algorithm on an unlabeled dataset, and it learns to find patterns or relationships within the data without specific guidance on the outputs.

## 7. Reinforcement Learning:

An AI learning paradigm where an agent learns by interacting with an environment. It receives feedback in the form of rewards or penalties based on its actions, allowing it to learn optimal behavior.

## 8. Natural Language Processing (NLP):

A branch of AI that enables machines to understand, interpret, and generate human language. NLP is used in applications such as language translation, chatbots, and sentiment analysis.

## 9. Computer Vision:

Focuses on enabling machines to interpret and make decisions based on visual data. Applications include image recognition, object detection, and facial recognition.

## 10. Expert Systems:

AI systems designed to emulate the decision-making ability of a human expert in a particular domain. They use rule-based systems and knowledge representation to solve specific problems.

## 11. Robotics:

The integration of AI into robotic systems to enable them to perceive their environment, make decisions, and carry out tasks autonomously. Robotic process automation (RPA) is a common application.

## 12. AI Ethics:

The study and implementation of ethical principles in the development and use of AI systems. This includes addressing issues like bias, transparency, accountability, and the societal impact of AI technologies.

## 13. AI Applications:

- **Healthcare:** Diagnosis, personalized medicine.
- **Finance:** Fraud detection, algorithmic trading.

- **Automotive:** Autonomous vehicles.

- **Entertainment:** Recommendation systems.

- **Security:** Facial recognition, threat detection.

## 14. Challenges in AI:

- **Ethical Concerns:** Bias in algorithms, privacy issues.

- **Explainability:** Understanding and interpreting complex AI models.

- **Data Quality:** AI systems are highly dependent on quality data.

- **Job Displacement:** Concerns about automation replacing certain jobs.

## 15. Future of AI:

Ongoing advancements, research, and applications in AI are expected to shape the future, with possibilities ranging from improved AI capabilities to ethical considerations and societal impacts.

Understanding the basics of AI provides a foundation for exploring its diverse applications and the ongoing developments in this rapidly evolving field.

## 3.1 Understanding Machine Learning

Machine Learning (ML) is a field of artificial intelligence that focuses on the development of algorithms and models that enable computers to learn from data and make decisions or predictions without being explicitly programmed.

The core idea behind machine learning is to create systems that can automatically learn and improve from experience.

1. **Key Concepts in Machine Learning:**

   a) **Learning from Data:**

   In traditional programming, humans write explicit rules for computers to follow. In machine learning, algorithms learn from patterns and data, adjusting their parameters to improve performance.

b) **Data:**

Data is the foundation of machine learning. Algorithms require labeled or unlabeled datasets to train, validate, and test their performance. These datasets consist of input features and corresponding output labels.

c) **Features and Labels:**

Features are the input variables or attributes used to make predictions, and labels are the output variables that the model aims to predict. The relationship between features and labels is learned during the training process.

d) **Training and Testing:**

The machine learning model is trained on a subset of the data, learning patterns and relationships. The trained model is then tested on new, unseen data to evaluate its performance and generalization capabilities.

2. **Types of Machine Learning:**

- **Supervised Learning:** The algorithm is trained on a labeled dataset, where it learns to map inputs to outputs. It makes predictions or classifications based on the learned patterns.

- **Unsupervised Learning:** The algorithm is given an unlabeled dataset and must find patterns or relationships within the data without specific guidance on the outputs.

- **Reinforcement Learning:** An agent learns to make decisions by interacting with an environment. It receives feedback in the form of rewards or penalties based on its actions.

3. **Algorithms:**

Machine learning algorithms are mathematical models that perform tasks such as classification, regression, clustering, or reinforcement learning. Common algorithms include decision trees, support vector machines, and neural networks.

4. **Model Evaluation:**

The performance of a machine learning model is assessed using various metrics such as accuracy, precision, recall, and F1 score. Cross-validation is often used to ensure robust evaluation.

**Overfitting and Underfitting:**

- **Overfitting:** Occurs when a model learns the training data too well but fails to generalize to new data.

- **Underfitting:** Occurs when a model is too simple to capture the underlying patterns in the data.

**Hyperparameters:**

Parameters of a machine learning model that are not learned from data but are set before the training process. Tuning hyperparameters is crucial for optimizing model performance.

5. **Ensemble Learning:**

Combining multiple machine learning models to improve overall performance and generalization. Ensemble methods include bagging, boosting, and stacking.

6. **Feature Engineering:**

The process of selecting, transforming, and creating features to enhance the performance of machine learning models.

**Applications of Machine Learning:**

- **Natural Language Processing (NLP):** Language translation, sentiment analysis, chatbots.

- **Computer Vision:** Image recognition, object detection, facial recognition.

- **Recommendation Systems:** Personalized content recommendations in streaming services, product recommendations in e-commerce.

- **Healthcare:** Diagnosis, personalized medicine, predictive analytics.

- **Finance:** Fraud detection, credit scoring, algorithmic trading.

- **Autonomous Vehicles:** Path planning, object detection, navigation.

**Challenges and Future Trends:**

- **Ethical Considerations:** Addressing bias, fairness, and transparency in machine learning models.

- **Explainability:** Understanding and interpreting complex models for better trust and accountability.

- **Advancements:** Ongoing research in deep learning, reinforcement learning, and the integration of AI into various industries.

Machine learning continues to evolve, with innovations and applications transforming industries and shaping the future of technology. Understanding the fundamental concepts provides a basis for exploring the complexities and possibilities within this dynamic field.

## 3.2 Machine Learning in Civil Engineering and Construction

Machine Learning (ML), a subset of artificial intelligence, has found diverse applications in civil engineering and construction, contributing to enhanced efficiency, decision-making, and overall project management.

Here are several ways in which machine learning is applied in the field:

1. **Predictive Analytics for Project Scheduling:**

    Machine learning algorithms analyze historical project data, including completion times, resource utilization, and external factors like weather conditions. This enables the prediction of project timelines, facilitating more accurate project scheduling.

2. **Cost Estimation and Budgeting:**

ML models process and analyze data on historical project costs, material prices, and labor expenses to provide accurate cost estimations. This helps in creating realistic budgets and minimizing the risk of cost overruns.

3. **Construction Equipment Optimization:**

Machine learning algorithms optimize the use of construction equipment by analyzing usage patterns, maintenance records, and performance data. This enables predictive maintenance, reducing downtime and extending the lifespan of equipment.

4. **Quality Control and Defect Detection:**

ML is employed for automated quality control by analyzing images or sensor data to detect defects in construction components. This ensures that the final product meets quality standards, and deviations are identified early in the construction process.

5. **Risk Management:**

ML models assess and predict project risks by analyzing historical data, project parameters, and external factors. This aids in developing proactive risk mitigation strategies, improving decision-making, and minimizing project uncertainties.

6. **Materials Management and Optimization:**

ML algorithms optimize the selection and management of construction materials by analyzing data on material properties, costs, and performance. This assists in choosing the most suitable materials for specific projects, considering factors like sustainability and durability.

7. **Automated Drawing Recognition:**

Machine learning is utilized for the automated recognition and interpretation of architectural and engineering drawings. This reduces manual effort in design review processes and ensures that construction aligns with specified plans and blueprints.

8. **Energy Efficiency in Building Design:**

ML contributes to energy-efficient building design by analyzing data on building orientation, materials, and climate conditions. This helps architects and engineers optimize designs for reduced energy consumption and improved sustainability.

9. **Construction Site Safety:**

ML applications enhance safety on construction sites by analyzing data from various sources, such as surveillance cameras and wearable devices. These systems can detect safety violations, monitor worker activities, and issue real-time alerts to prevent accidents.

10. **Geotechnical Analysis:**

Machine learning assists in geotechnical analysis by processing data related to soil properties, geological conditions, and site history. This information is crucial for site suitability assessments and designing foundations based on geological data.

11. **Construction Project Management:**

ML is employed in project management for tasks such as resource allocation, task scheduling, and progress tracking. This enables project managers to make data-driven decisions, improving overall project efficiency.

12. **Automated Inspection and Monitoring:**

ML-powered drones or robotic systems can be used for automated inspection and monitoring of construction sites, structures, and infrastructure. This enhances the speed and accuracy of inspections, ensuring early detection of potential issues.

In summary, the integration of machine learning in civil engineering and construction offers a wide array of benefits, ranging from predictive analytics and cost estimation to quality control and safety enhancements. As technology continues to advance, machine learning is poised to play an increasingly vital role in shaping the future of the construction industry.

## 3.3 Understanding Neural Networks

Neural networks constitute a foundational element within the realm of artificial intelligence (AI), drawing inspiration from the intricate workings of the human brain.

This computational paradigm involves interconnected nodes or artificial neurons, organized into layers, providing the capacity to decipher intricate patterns and relationships within data. The architecture typically includes an input layer for data reception, hidden layers for data processing, and an output layer for result generation. The training process involves adjusting the weights of connections between neurons based on labeled datasets, enabling the network to learn and adapt, making it adept at tasks such as image and speech recognition, natural language processing, and decision-making.

1. **Types of Neural Networks:**

   Neural networks come in various types, each designed for specific applications and tasks. Feedforward neural networks, the most common type, transmit data in one direction from the input to the output layer. Recurrent neural networks (RNNs) introduce feedback loops, enabling them to consider previous information in their decision-making process. Convolutional neural networks (CNNs) are tailored for image processing tasks, using convolutional layers to extract features from visual data. Long Short-Term Memory networks (LSTMs) address the limitations of traditional RNNs in handling long-term dependencies in data.

2. **Deep Learning and Neural Networks:**

   Deep learning, a subfield of machine learning, is closely intertwined with neural networks. It involves the utilization of deep neural networks, which are characterized by multiple hidden layers. The depth of these networks allows them to learn hierarchical representations of data, enhancing their ability to grasp intricate features and nuances. Deep learning has exhibited remarkable success in tasks such as image and speech recognition, demonstrating its prowess in complex problem-solving.

3. **Training Neural Networks:**

   The training process of neural networks involves exposing the network to labeled datasets, allowing it to adjust its internal parameters, or weights, through a process called backpropagation. This iterative learning mechanism fine-tunes the network's ability to make accurate predictions or classifications. Techniques such as dropout and regularization are employed to prevent overfitting and improve generalization.

4. **Applications of Neural Networks:**

   Neural networks find applications across diverse fields. In computer vision, they excel at image recognition, object detection, and facial recognition. In natural language processing, they power language translation, chatbots, and sentiment analysis. Neural networks are also integral in healthcare for tasks like disease diagnosis and personalized medicine. Additionally, they play a role in financial sectors for fraud detection and algorithmic trading, contribute to the development of autonomous vehicles in the automotive industry, enhance entertainment through recommendation systems, and bolster security measures with technologies like facial recognition and threat detection.

5. **Challenges and Future of Neural Networks:**

Despite their effectiveness, neural networks face challenges such as interpretability, the "black box" nature of complex models, and the need for large datasets for training. The future of neural networks holds promise with ongoing advancements, including improved architectures, algorithms, and ethical considerations. As research continues to push the boundaries, neural networks are expected to evolve and contribute significantly to the ever-expanding landscape of artificial intelligence. Understanding the nuances of neural networks provides a solid foundation for navigating their applications and developments in this dynamic and rapidly evolving field.

6. **Applications:**
   a. **Image and Speech Recognition:** Convolutional neural networks are widely used for image recognition, and recurrent neural networks for speech recognition.
   b. **Natural Language Processing (NLP):** Neural networks power language translation, sentiment analysis, and chatbots.

Neural networks are a foundational technology in the field of artificial intelligence, driving innovations and advancements across various domains. As they continue to evolve, neural networks contribute to the development of intelligent systems with improved capabilities and broader applications.

## 3.4 Neural Networks in Civil Engineering and Construction

Neural networks, a subset of artificial intelligence (AI), have found applications in various aspects of civil engineering and construction, contributing to improved efficiency, accuracy, and decision-making.

Here are several ways in which neural networks are applied in the field:

1. **Structural Health Monitoring:**

   Neural networks are utilized in the analysis of sensor data for structural health monitoring. They can predict and identify potential issues, such as fatigue, deformation, or damage in buildings and infrastructure, aiding in predictive maintenance and ensuring structural integrity.

2. **Predictive Analytics for Project Scheduling:**

Neural networks are employed to analyze historical project data, weather conditions, and other relevant factors to predict project timelines accurately. This aids project managers in creating more realistic schedules and anticipating potential delays.

3. **Construction Material Optimization:**

Neural networks help optimize the selection and usage of construction materials. By analyzing data on material properties, cost, and performance, these networks can recommend the most suitable materials for specific construction projects, considering factors like durability and sustainability.

4. **Cost Estimation and Budgeting:**

Neural networks contribute to accurate cost estimation and budgeting by analyzing historical project data, material costs, labor expenses, and other relevant factors. This assists in creating more precise project budgets and minimizing the risk of cost overruns.

5. **Construction Equipment Management:**

Neural networks are applied to monitor and manage construction equipment. By analyzing usage patterns, performance data, and maintenance records, these networks help predict equipment failures, optimize maintenance schedules, and improve overall equipment efficiency.

6. **Site Safety Monitoring:**

Neural networks enhance safety on construction sites by analyzing data from surveillance cameras, wearable devices, and other sensors. They can detect potential safety hazards, monitor workers' activities, and issue real-time alerts to prevent accidents.

7. **Automated Quality Control:**

Neural networks contribute to automated quality control processes by analyzing images of construction components. They can identify defects, deviations from design specifications, and construction errors, ensuring that the final product meets quality standards.

8. **Energy Efficiency in Building Design:**

Neural networks are used in the design phase to optimize energy efficiency in buildings. They analyze factors such as building orientation, materials, and climate data to recommend design modifications that reduce energy consumption and enhance sustainability.

9. **Construction Site Layout Planning:**

Neural networks assist in optimizing construction site layout planning by considering factors like material storage, equipment placement, and worker movements. This helps minimize delays, enhance workflow efficiency, and improve overall site productivity.

10. **Geotechnical Analysis and Site Suitability:**

Neural networks aid in geotechnical analysis by processing data related to soil properties, geological conditions, and site history. This information is crucial for assessing the suitability of a site for construction and designing foundations accordingly.

11. **Automated Drawing Recognition and Interpretation:**

Neural networks are employed to recognize and interpret architectural and engineering drawings. This assists in automating tasks related to design review, ensuring that construction adheres to specified plans and blueprints.

In conclusion, the integration of neural networks in civil engineering and construction contributes to advancements in various aspects of project planning, execution, and management. As technology continues to evolve, neural networks are likely to play an increasingly significant role in shaping the future of the construction industry.

## 3.5 Understanding Natural Language Processing

Natural Language Processing (NLP) is a branch of artificial intelligence (AI) that focuses on the interaction between computers and humans using natural                                                   language.

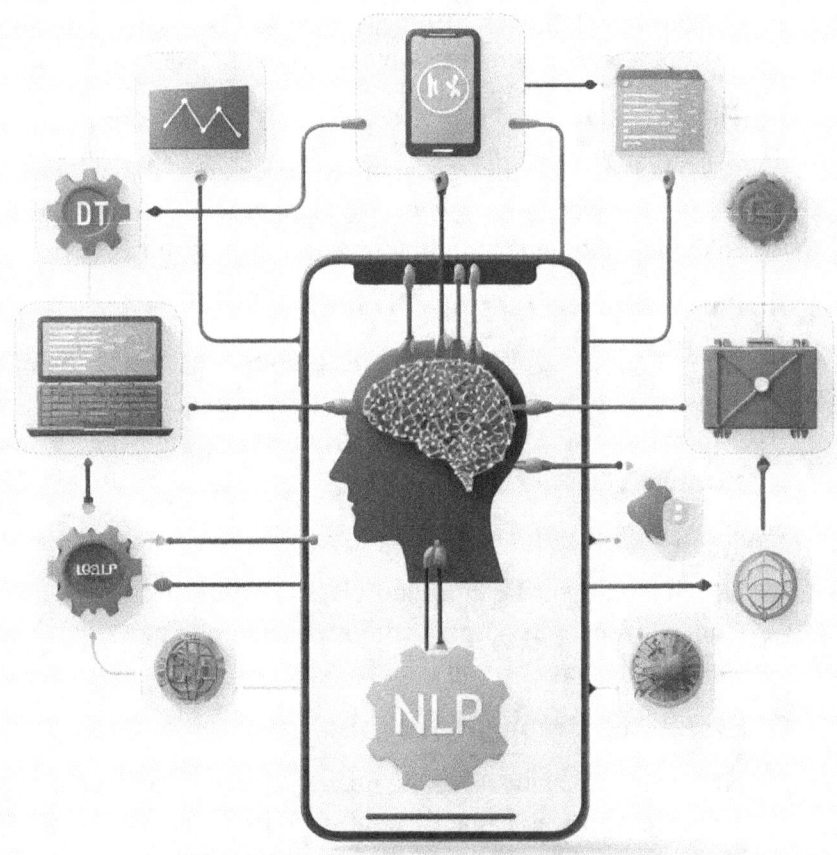

The goal of NLP is to enable computers to understand, interpret, and generate human language in a way that is both meaningful and contextually relevant. NLP plays a crucial role in applications such as language translation, sentiment analysis, chatbots, and text summarization. Here's an overview of key concepts in NLP:

**Key Concepts:**

1. **Natural Language Understanding (NLU):**

NLU is the ability of a computer system to comprehend and interpret human language. It involves tasks such as parsing, semantic analysis, and extracting meaning from text.

2. **Natural Language Generation (NLG):**

NLG is the process of generating meaningful and coherent human-like language from structured data. This can include tasks like text summarization and content creation.

3. **Tokenization:**

Tokenization is the process of breaking down text into individual words or tokens. It is a crucial step in NLP for analyzing and understanding the structure of sentences.

4. **Part-of-Speech Tagging (POS):**

POS tagging involves assigning grammatical categories (such as noun, verb, adjective) to each word in a sentence. This information is valuable for understanding the syntactic structure of a sentence.

5. **Named Entity Recognition (NER):**

NER is the identification and classification of entities, such as names of people, organizations, locations, dates, and other specific elements within a text.

6. **Syntax and Grammar Analysis:**

NLP systems analyze the syntactic structure and grammar of sentences to understand relationships between words and phrases, enabling them to derive meaning.

7. **Semantic Analysis:**

Semantic analysis focuses on understanding the meaning of words and how they relate to each other in context. It involves interpreting the intended meaning of a sentence beyond its literal interpretation.

8. **Sentiment Analysis:**

Sentiment analysis, or opinion mining, involves determining the sentiment expressed in a piece of text. It is commonly used to assess whether a text is positive, negative, or neutral.

9. **Machine Translation:**

NLP is integral to machine translation systems, enabling computers to automatically translate text or speech from one language to another.

10. **Chatbots and Virtual Assistants:**

NLP is employed in chatbots and virtual assistants to understand user queries and respond in a natural and conversational manner.

## Applications of Natural Language Processing:

- **Search Engines:** NLP improves search engine understanding, leading to more accurate results based on user queries.

- **Speech Recognition:** NLP powers systems that convert spoken language into written text, facilitating voice commands and dictation.

- **Email Filtering:** NLP is used in spam filters to identify and filter out unwanted emails.

- **Text Summarization:** NLP systems can generate concise summaries of lengthy texts, making information more accessible.

- **Language Translation:** NLP enables automatic translation between languages, enhancing cross-cultural communication.

## Challenges and Considerations:

- **Ambiguity:** Natural language often contains ambiguity, and understanding context is challenging.

- **Language Variations:** Diverse language use, including slang, regional variations, and colloquial expressions, poses challenges for NLP systems.

- **Context Understanding:** Grasping the context and intent behind a statement requires a deep understanding of the topic.

## Future Trends in Natural Language Processing:

- **BERT and Transformer Models:** Transformer-based models, such as BERT (Bidirectional Encoder Representations from

Transformers), have shown remarkable performance improvements in various NLP tasks.

- **Multimodal NLP:** Integration of NLP with other modalities, such as images and videos, for a more comprehensive understanding of content.

- **Conversational AI:** Advancements in creating more natural and context-aware interactions in conversational agents.

Natural Language Processing continues to evolve, driven by ongoing research and advancements in AI technologies. As NLP systems become more sophisticated, they have the potential to enhance human-computer interactions and enable machines to better understand and respond to natural language input.

## 3.6 Natural Language Processing in Civil Engineering and Construction

Natural Language Processing (NLP) in civil engineering and construction involves the application of computational linguistics and machine learning techniques to understand and process human language.

This technology offers several benefits in the industry, streamlining communication, enhancing documentation processes, and improving overall project management. Here are ways in which NLP is applied in civil engineering and construction:

### 12. Document Analysis and Summarization:

NLP algorithms can analyze and summarize large volumes of construction-related documents, including contracts, reports, and specifications. This helps professionals quickly extract key information, identify critical project requirements, and make informed decisions.

### 13. Automated Bidding and Proposal Analysis:

NLP can automate the analysis of bidding documents and proposals. By extracting and understanding information from textual data, it assists in evaluating bids, comparing proposals, and identifying relevant terms and conditions.

### 14. Risk Assessment and Contract Management:

NLP applications aid in the identification and analysis of potential risks in construction contracts. By processing contract language, NLP can highlight clauses, conditions, or terms that may pose risks, helping project managers make more informed decisions.

### 15. Information Extraction from Emails and Communications:

NLP algorithms can parse through emails, project communications, and other textual data to extract essential information. This can include updates on project status, change orders, and important announcements, improving communication efficiency.

### 16. Automated Drawing and Specification Analysis:

NLP can assist in the analysis of architectural drawings and specifications. By understanding the natural language descriptions within these documents, NLP systems can identify key features, requirements, and constraints, facilitating better project planning.

### 17. Chatbots for Project Communication:

Chatbots powered by NLP can enhance communication on construction projects. They provide instant responses to queries related to project timelines, material specifications, and other relevant information, improving accessibility and responsiveness.

### 18. Facilitating Data Entry and Retrieval:

NLP simplifies data entry processes by allowing project teams to input information in natural language, reducing the need for manual data entry. It also enables users to retrieve specific information from databases or project documentation using natural language queries.

### 19. Translation Services for Multilingual Projects:

In projects involving teams from diverse linguistic backgrounds, NLP-based translation services can be employed to facilitate seamless communication. This ensures that all team members can understand and contribute effectively, reducing the risk of miscommunication.

### 20. Enhancing Building Information Modeling (BIM):

NLP can be integrated into BIM systems to enable users to interact with the model using natural language queries. This makes it easier for non-technical stakeholders to access and understand the information contained in the BIM model.

### 21. Regulatory Compliance and Code Analysis:

NLP applications can assist in analyzing and interpreting building codes and regulatory documents. By processing the language within these documents, NLP systems can help ensure that construction projects comply with relevant regulations.

In summary, NLP is a valuable tool in civil engineering and construction, offering solutions to challenges related to document analysis, communication, and project management. As technology continues to advance, the integration of NLP is likely to become more widespread, contributing to increased efficiency and effectiveness in the industry.

## 3.7 Understanding Computer Vision

Computer Vision is a field of artificial intelligence that empowers machines to interpret and make decisions based on visual data.

It involves developing algorithms and systems that enable computers to understand and extract information from images or videos in a way that simulates human vision. The applications of computer vision range from image recognition and object detection to facial recognition and autonomous vehicles.

**1. Key Concepts in Computer Vision:**

- **Image Acquisition:**

The process of obtaining visual data is the first step in computer vision. This can be through cameras, sensors, or other imaging devices.

- **Image Processing:**

  Image processing involves manipulating and enhancing visual data to improve its quality or extract relevant features. Operations include filtering, noise reduction, and image enhancement.

- **Feature Extraction:**

  Feature extraction involves identifying key patterns or characteristics in the visual data, such as edges, corners, or textures. These features are crucial for subsequent analysis.

- **Image Recognition:**

  Image recognition is the task of identifying and categorizing objects or patterns within an image. Deep learning models, such as convolutional neural networks (CNNs), excel at image recognition tasks.

- **Object Detection:**

  Object detection goes beyond recognition by locating and identifying multiple objects within an image or video. It involves bounding box prediction around objects of interest.

- **Semantic Segmentation:**

  Semantic segmentation assigns a class label to each pixel in an image, enabling the understanding of the specific regions and boundaries of objects.

- **Motion Analysis:**

  Motion analysis involves tracking objects or understanding the movement of objects within a sequence of frames. It is crucial for applications like video surveillance and autonomous vehicles.

- **3D Reconstruction:**

3D reconstruction aims to create a three-dimensional representation of a scene or object from two-dimensional visual data, providing depth information.

- **Facial Recognition:**

   Facial recognition involves identifying and verifying individuals based on their facial features. It has applications in security, authentication, and personalization.

- **Image Understanding:**

   Image understanding combines various computer vision tasks to comprehend the content, context, and relationships within visual data.

## 2. Applications of Computer Vision:

- **Autonomous Vehicles:** Computer vision is essential for tasks such as lane detection, object recognition, and pedestrian tracking in autonomous vehicles.

- **Medical Image Analysis:** Identifying and analyzing patterns in medical images for diagnosis and treatment planning.

- **Augmented Reality (AR) and Virtual Reality (VR):** Enhancing user experiences by overlaying digital information onto the real world or creating immersive virtual environments.

- **Industrial Automation:** Quality control, defect detection, and robotic guidance in manufacturing processes.

- **Retail:** Shelf monitoring, customer behavior analysis, and cashier-free checkout systems.

- **Security and Surveillance:** Facial recognition, object tracking, and anomaly detection in video surveillance.

## 3. Challenges and Considerations:

- **Data Variability:** Handling diverse and complex visual data with variations in lighting, viewpoint, and object appearance.

- **Interpretable Models:** Making computer vision models more interpretable and explainable for trust and accountability.

- **Ethical Concerns:** Addressing issues related to privacy, bias, and fairness in applications like facial recognition.

4. **Future Trends in Computer Vision:**

  - **Generative Adversarial Networks (GANs):** GANs are being used for tasks like image synthesis, style transfer, and data augmentation.

  - **Edge Computing:** Performing computer vision tasks closer to the source of data for real-time processing.

  - **Human Pose Estimation:** Advancements in understanding and predicting the poses of humans in images or videos.

Computer Vision continues to advance rapidly, driven by innovations in deep learning, hardware, and real-world applications. As technology evolves, computer vision is expected to play an increasingly significant role in transforming industries and enhancing our interaction with the visual world.

# 3.8 Computer vision in civil engineering and construction

Computer vision plays a significant role in civil engineering and construction, offering innovative solutions to enhance efficiency, safety, and accuracy in various aspects of the industry.

Here are several ways in which computer vision is applied in civil engineering and construction:

1. **Project Monitoring and Management:**

   Computer vision enables real-time monitoring of construction sites through the analysis of images and videos. Drones equipped with cameras can capture aerial footage, providing project managers with a comprehensive view of the site. This aids in progress tracking, resource allocation, and early identification of potential issues.

2. **Site Inspection and Surveys:**

Computer vision assists in automating site inspections and surveys. Drones or ground-based cameras can capture high-resolution images, which can be processed to generate 3D models of the construction site. This streamlines the surveying process and provides accurate data for design and planning.

3. **Safety Monitoring:**

Computer vision systems can be implemented for safety monitoring on construction sites. This includes the detection of safety violations, such as workers not wearing appropriate safety gear or the identification of potential hazards. Automated systems can issue alerts in real-time to prevent accidents.

4. **Quality Control:**

Computer vision helps in quality control by analyzing images of construction components and structures. It can detect defects, deviations from design specifications, or construction errors early in the process, allowing for corrective measures to be implemented promptly.

5. **Facial Recognition for Access Control:**

Facial recognition technology can be integrated into access control systems on construction sites. This enhances security by allowing only authorized personnel to enter specific areas. It also aids in tracking attendance and ensures that workers have the required certifications.

6. **Construction Equipment Monitoring:**

Computer vision is utilized to monitor the usage and condition of construction equipment. By analyzing images or video feeds, it can assess the performance and maintenance needs of machinery, helping to prevent breakdowns and optimize equipment utilization.

7. **Augmented Reality (AR) for Design Visualization:**

Computer vision, combined with augmented reality, allows engineers and architects to visualize construction designs in the real world. This aids in on-site decision-making, as stakeholders can see how the finished project will look and identify any potential clashes or design issues.

## 8. Materials Management:

Computer vision helps in automating the tracking and management of construction materials. By analyzing images or RFID tags, it can monitor the arrival, usage, and movement of materials on the site, improving inventory management and reducing the risk of shortages or overstock.

## 9. Automated Construction:

Advances in computer vision contribute to the development of autonomous construction equipment. Automated vehicles and robots equipped with computer vision systems can perform tasks such as excavation, bricklaying, and concrete pouring, enhancing efficiency and reducing labor requirements.

In summary, computer vision applications in civil engineering and construction are diverse, ranging from project monitoring and safety to quality control and automation. As technology continues to evolve, these applications are likely to become even more integral to the industry, providing innovative solutions to age-old challenges.

# 3.9 AI in Decision Making

Artificial Intelligence (AI) has significantly impacted decision-making processes across various industries by providing data-driven insights, automating tasks, and enhancing overall efficiency.

Here are key aspects of AI in decision making:

1.  **Data Analysis and Pattern Recognition:**

    AI systems excel at processing vast amounts of data quickly. They analyze historical data to identify patterns, trends, and correlations, providing valuable insights for decision-making.

2.  **Predictive Analytics:**

AI models, particularly machine learning algorithms, can predict future outcomes based on historical data. This is valuable for forecasting trends, demand, and potential risks, aiding decision-makers in planning and strategy.

3. **Automation of Routine Decisions:**

AI automates routine and repetitive decision-making processes. This allows human decision-makers to focus on more complex and strategic aspects of their roles.

4. **Personalized Recommendations:**

In industries like e-commerce, entertainment, and marketing, AI algorithms analyze user behavior to provide personalized recommendations. This enhances user experience and supports decision-making related to product choices or content consumption.

5. **Risk Management:**

AI is employed in risk assessment and management. It can analyze various factors to identify potential risks and suggest strategies for mitigation, particularly in finance, insurance, and project management.

6. **Natural Language Processing (NLP) for Decision Support:**

NLP enables AI systems to understand and process human language. This is valuable for extracting information from unstructured data sources, such as textual documents, to support decision-making.

7. **Supply Chain Optimization:**

AI optimizes supply chain decisions by analyzing factors like demand forecasting, inventory levels, and transportation logistics. This ensures efficient resource allocation and reduces operational costs.

8. **Healthcare Diagnosis and Treatment Plans:**

AI in healthcare assists in diagnosing medical conditions and recommending personalized treatment plans based on patient data. This supports medical professionals in making more informed decisions.

9. **Algorithmic Trading in Finance:**

AI-driven algorithms analyze market data, predict trends, and execute trades at high speeds. This enhances decision-making in financial markets.

10. **Human Resource Management:**

AI is used in recruitment processes to analyze resumes, screen candidates, and even conduct initial interviews. This streamlines the hiring process and supports decisions related to candidate selection.

11. **Smart Cities and Urban Planning:**

AI aids in decision-making for urban planners by analyzing data from various sources, including traffic patterns, energy consumption, and citizen behavior. This information informs decisions related to infrastructure development and city management.

12. **Customer Service and Chatbots:**

AI-powered chatbots assist in decision-making by providing instant responses to customer queries, handling routine tasks, and escalating complex issues to human agents.

13. **Quality Control in Manufacturing:**

AI systems analyze production data and identify defects or deviations from quality standards. This supports decisions related to improving manufacturing processes and product quality.

14. **Legal and Contract Analysis:**

AI tools can review and analyze legal documents, contracts, and case law. This aids legal professionals in decision-making processes related to legal research and risk assessment.

15. **Real-Time Decision Support:**

AI systems can process real-time data and provide immediate insights. This is valuable in scenarios where timely decisions are critical, such as in emergency response or cybersecurity.

**Challenges and Considerations:**

- **Interpretability and Explainability:** Understanding how AI models reach specific decisions is crucial for trust and accountability.

- **Ethical Considerations:** AI systems must be designed and used ethically to avoid bias, discrimination, and unintended consequences.

- **Data Privacy:** Handling sensitive information requires robust measures to ensure privacy and compliance with regulations.

**Future Trends:**

- **Explainable AI (XAI):** Advancements in making AI models more interpretable and explainable.

- **AI Collaboration:** Increased collaboration between AI systems and human decision-makers, fostering a more synergistic approach.

- **Continuous Learning Models:** AI systems that can adapt and learn continuously, improving decision-making capabilities over time.

In summary, AI is revolutionizing decision-making across diverse domains, providing valuable insights, automating routine tasks, and enabling more informed and efficient choices. However, careful consideration of ethical and interpretability issues is crucial for responsible AI deployment.

## 3.10 AI in Decision Making for Civil Engineering and Construction

Artificial Intelligence (AI) plays a significant role in decision-making processes within the field of Civil Engineering and Construction. Here are several ways in which AI is applied to enhance decision-making in these domains:

1. **Project Planning and Scheduling:**

   AI algorithms can analyze historical project data to optimize project planning and scheduling.

   Machine learning models can predict potential delays, resource constraints, and suggest adjustments to the project timeline for better efficiency.

2. **Risk Assessment and Management:**

   AI tools can assess and predict potential risks in construction projects based on historical data, environmental factors, and project-specific variables.

   Risk management systems powered by AI can provide real-time risk assessments, allowing project managers to make informed decisions to mitigate potential issues.

3. **Cost Estimation and Budgeting:**

   AI can assist in accurate cost estimation by analyzing historical cost data and considering various project parameters.

   Machine learning models can predict potential cost overruns and provide recommendations for budget adjustments.

4. **Design Optimization:**

   AI algorithms can optimize designs by analyzing various parameters such as structural integrity, material efficiency, and environmental impact.

   Generative design, a subset of AI, can explore numerous design alternatives and propose optimal solutions based on specified criteria.

5. **Supply Chain Management:**

AI facilitates efficient supply chain management by predicting material demands, optimizing inventory levels, and identifying potential bottlenecks.

Predictive analytics can help in anticipating material shortages or delays in the supply chain, enabling proactive decision-making.

6. **Quality Control and Assurance:**

AI-powered computer vision systems can perform real-time quality inspections of construction sites and materials.

Machine learning algorithms can identify defects, deviations from design specifications, and potential issues, allowing for prompt corrective actions.

7. **Equipment and Resource Optimization:**

AI can optimize the utilization of construction equipment by analyzing usage patterns and predicting maintenance needs.

Smart scheduling algorithms can allocate resources efficiently, minimizing downtime and enhancing overall project productivity.

8. **Energy Efficiency and Sustainability:**

AI can contribute to designing energy-efficient buildings and infrastructure by analyzing environmental data and recommending sustainable practices.

Decision support systems can assist in selecting materials and construction methods with lower environmental impact.

9. **Safety Monitoring and Compliance:**

AI-powered sensors and cameras can monitor construction sites for safety compliance.

Machine learning algorithms can analyze data to identify potential safety hazards and alert project managers to take preventive measures.

10. **Data-Driven Insights:**

AI enables the analysis of large datasets, extracting meaningful insights that aid in decision-making.

Predictive analytics can help anticipate challenges, trends, and opportunities, allowing project managers to make informed decisions.

## 11. Autonomous Construction Vehicles:

AI-driven autonomous vehicles and drones can assist in tasks such as surveying, site inspection, and material transportation, improving efficiency and safety.

## 12. Regulatory Compliance:

AI can help in tracking and ensuring compliance with local building codes, regulations, and standards.

Automated compliance checks can reduce the risk of legal issues and ensure that projects meet the required standards.

Implementing AI in decision-making processes in civil engineering and construction can lead to more efficient project management, cost savings, improved safety, and sustainable practices. However, it's essential to integrate AI technologies in collaboration with domain experts and adhere to ethical considerations and regulatory requirements.

## 3.11 Real-world Examples / Use Cases

1. **AI in Civil Engineering and Construction:**

   i. **Project Risk Management (Company: nPlan):**

      - nPlan employs AI to predict and manage project risks in construction. The platform analyzes historical project data to forecast potential risks and uncertainties, enhancing decision-making.

   ii. **Construction Automation (Company: Built Robotics):**

      - Built Robotics utilizes AI for the automation of heavy construction equipment. Their AI-driven systems retrofit standard construction machinery, enabling autonomous operation for tasks like excavation and grading.

2. **Machine Learning in Civil Engineering and Construction:**

   iii. **Construction Project Scheduling (Company: Assignar):**

      - Assignar employs machine learning for construction project scheduling. The platform uses historical project data to optimize scheduling, allocate resources efficiently, and reduce delays.

   iv. **Material Strength Prediction (Company: DeepMind - Alphabet):**

      - DeepMind, a subsidiary of Alphabet, applies machine learning to predict the structural behavior of materials. Their research explores how ML algorithms can improve material strength predictions for construction applications.

3. **Neural Networks in Civil Engineering and Construction:**

   v. **Structural Health Monitoring (Company: COWI):**

      - COWI utilizes neural networks for structural health monitoring of bridges. The company employs advanced algorithms to analyze sensor data and detect potential structural issues in real-time.

   vi. **Optimized Building Design (Company: Thornton Tomasetti):**

- Thornton Tomasetti incorporates neural networks for optimized building design. Their AI-driven tools explore various design possibilities, considering factors like sustainability, cost, and structural efficiency.

4. **Natural Language Processing (NLP) in Civil Engineering and Construction:**

vii. **Contract Analysis (Company: Leverton - MRI Software):**

- Leverton, now a part of MRI Software, uses NLP for contract analysis in real estate and construction. The platform extracts key information from contracts, streamlining document review processes.

viii. **Project Communication (Company: Procore):**

- Procore integrates NLP into its project management platform for construction. The NLP capabilities facilitate seamless communication by extracting insights from project-related documents and discussions.

5. **Computer Vision in Civil Engineering and Construction:**

ix. **Construction Site Monitoring (Company: OpenSpace):**

- OpenSpace employs computer vision for construction site monitoring. Their platform uses 360-degree cameras and AI to create a visual record of construction progress, aiding in project management.

x. **Defect Detection (Company: SmartVid):**

- SmartVid utilizes computer vision for defect detection in construction. The platform analyzes images and videos from construction sites to identify potential issues and ensure quality control.

6. **AI in Decision Making in Civil Engineering and Construction:**

xi. **Bid Analysis (Company: Bidgely):**

- Bidgely applies AI for bid analysis in construction projects. The platform uses machine learning algorithms to assess bids, providing insights for more informed decision-making.

xii. **Resource Allocation (Company: Assignar):**

- Assignar, in addition to project scheduling, employs AI for resource allocation in construction. The platform optimizes the deployment of resources based on historical and real-time data, improving overall efficiency.

## 3.12 Chapter Summary: Key Points

1. AI simulates human intelligence in machines, involving algorithms for learning, problem-solving, and decision-making.

2. Narrow AI (Weak AI) is task-specific, while General AI (Strong AI) emulates broad human intelligence (theoretical).

3. ML, a subset of AI, improves performance through data-driven learning, including supervised, unsupervised, and reinforcement learning.

4. Deep learning, inspired by neural networks, involves multi-layered networks for complex decision-making and pattern recognition.

5. NLP enables machines to understand, interpret, and generate human language, applied in translation and chatbots.

6. Computer Vision uses algorithms for visual data interpretation, including image recognition and facial detection.

7. Expert Systems mimic human decision-making, using rule-based reasoning in applications like healthcare and finance.

8. The Turing Test assesses AI by its human-like conversational ability, aiming for indistinguishable interaction.

9. AI impacts healthcare diagnostics, fraud detection in finance, autonomous vehicles, and personalized recommendations.

10. Addressing bias, transparency, and societal impact is crucial for responsible AI development and deployment.

11. NLP in civil engineering streamlines communication and enhances project management by analyzing and summarizing construction-related documents.

12. Automated bidding and proposal analysis using NLP facilitates the evaluation of bids, comparison of proposals, and identification of relevant terms and conditions.

13. NLP applications contribute to risk assessment in construction contracts by identifying potential risks through the analysis of contract language.

14. NLP algorithms parse through emails and project communications, extracting essential information to improve communication efficiency in construction projects.

15. Automated drawing and specification analysis with NLP aids in understanding architectural documents, identifying key features, requirements, and constraints for better project planning.

16. Chatbots powered by NLP enhance communication on construction projects, providing instant responses to queries related to timelines and material specifications.

17. NLP simplifies data entry processes in construction projects, allowing natural language input and retrieval of specific information from databases.

18. Computer vision in civil engineering monitors projects in real-time, automates site inspections, ensures safety, and enhances quality control through image analysis and 3D modeling.

## 3.13 Concept Check: Q&A Sessions

1. What is the primary goal of Natural Language Processing (NLP) in the field of artificial intelligence?

2. How does Supervised Learning differ from Unsupervised Learning in the context of Machine Learning?

3. What are the key challenges associated with Neural Networks in artificial intelligence?

4. How does Machine Learning contribute to the field of civil engineering, specifically in construction projects?

5. What ethical considerations are crucial in the development and deployment of Artificial Intelligence (AI) technologies?

6. How does Natural Language Processing (NLP) contribute to the advancement of search engines?

7. How does Natural Language Processing (NLP) contribute to risk assessment in construction contracts?

8. What role does Computer Vision play in quality control within the construction industry?

9. In what ways does Artificial Intelligence (AI) impact decision-making in the field of healthcare?

10. How can Natural Language Processing (NLP) be utilized for communication efficiency in construction projects?

11. What is the significance of Explainable AI (XAI) in the context of decision-making, and why is it important?

12. How does Computer Vision contribute to safety monitoring on construction sites?

# 4. Applications of AI in Project Planning

Artificial Intelligence (AI) plays a crucial role in project planning across various industries, streamlining processes and enhancing efficiency.

In general, AI contributes to project planning by automating repetitive tasks, providing data-driven insights, and optimizing resource allocation. In civil engineering construction projects, AI applications further extend to complex tasks specific to the industry.

In project planning, AI algorithms analyze historical project data to identify patterns and trends, aiding in accurate project timelines and resource estimations. This data-driven approach enhances decision-making, enabling project managers to allocate resources efficiently and reduce the risk of delays.

In civil engineering, AI applications include structural analysis and design optimization. AI algorithms can assess various design parameters to recommend the most efficient and cost-effective structural solutions, contributing to optimal project planning. In the realm of project planning, Artificial Intelligence (AI) introduces a transformative impact through six key functions. Below diagram illustrates 6 key functions of AI in Project Planning.

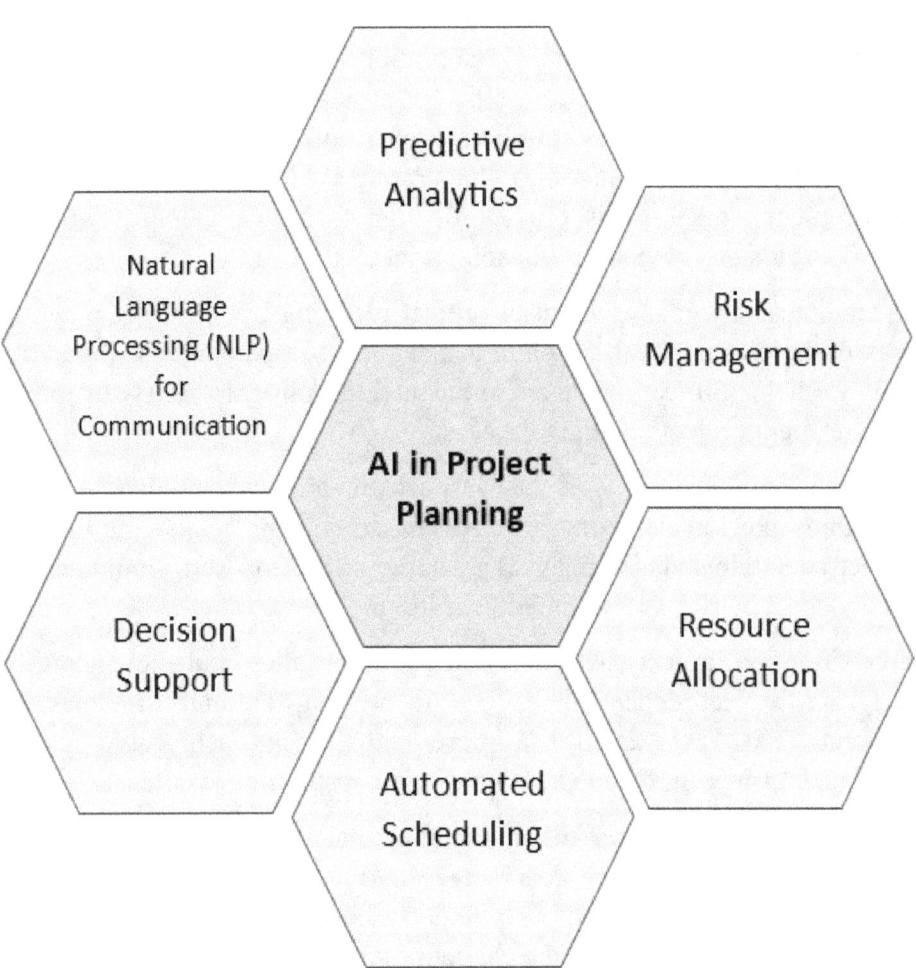

Predictive analytics, powered by AI algorithms, enhances accuracy in estimating project timelines, resource requirements, and potential risks by analyzing historical data and patterns. AI's role in risk management involves proactive identification of potential issues through the analysis of historical project data, enabling project managers to implement mitigation strategies and contingency plans. Resource allocation is optimized by AI, which analyzes project requirements, team capabilities, and resource availability, ensuring efficient workforce distribution. Automated scheduling, driven by AI tools, streamlines the scheduling process, reduces manual effort, and improves adherence to project timelines. Decision support from AI provides real-time insights, aiding project managers in faster and more informed decision-making, particularly in complex scenarios. Additionally, Natural Language Processing (NLP) tools powered by AI facilitate communication within project teams, automating routine inquiries and enhancing collaboration. These functions collectively contribute to more effective and efficient project planning, enabling data-driven decision-making, risk mitigation, and optimized resource utilization for successful project outcomes. Integrating AI seamlessly into project management workflows is crucial to align these technologies with the specific needs and objectives of each project.

AI facilitates risk management in project planning by assessing potential risks based on historical data and project-specific parameters. Predictive analytics helps project managers anticipate and mitigate risks, ensuring a more resilient project planning process.

Resource optimization is a key aspect of project planning, and AI algorithms excel in allocating resources based on demand, availability, and project constraints. This enhances resource utilization and minimizes the likelihood of bottlenecks or underutilization.

In construction projects, AI contributes to scheduling and sequencing by analyzing project dependencies and identifying optimal timelines for different tasks. This ensures a more realistic and achievable project timeline, considering the intricate nature of construction processes.

Machine learning models in project planning continuously learn from ongoing projects, adapting to changes and optimizing future planning processes. This adaptability is particularly beneficial in dynamic project environments, where unforeseen challenges may arise.

AI-powered tools assist in cost estimation by analyzing historical cost data and considering various project parameters. This helps project managers create more accurate and realistic budgets, reducing the likelihood of cost overruns.

Construction projects involve vast amounts of data, including design documents, specifications, and construction drawings. AI-based document management systems organize and extract relevant information, streamlining the documentation process in project planning.

AI enhances collaboration in project planning by providing a centralized platform for stakeholders to access real-time project data. This fosters communication, reduces errors, and ensures that all team members are working with up-to-date information.

In construction projects, AI applications extend to monitoring and controlling project progress. AI algorithms analyze on-site sensor data, satellite imagery, and other sources to provide real-time insights, enabling project managers to make informed decisions and adapt the project plan as needed.

In summary, AI's applications in project planning encompass data analysis, risk management, resource optimization, and collaboration, contributing to more efficient and effective planning processes. In the context of civil engineering construction projects, AI's role extends to structural design optimization, document management, and real-time progress monitoring, enhancing the overall project management lifecycle.

## 4.1 Predictive Analytics

Artificial Intelligence (AI) plays a significant role in predictive analytics within the realm of civil engineering and construction projects.

By leveraging advanced algorithms and machine learning techniques, AI enhances the ability to forecast outcomes, identify patterns, and optimize decision-making processes. Here are key applications of AI for predictive analytics in civil engineering and construction projects:

1. **Project Timeline Prediction:**

   AI algorithms analyze historical project data, including timelines, delays, and dependencies, to predict future project timelines accurately. This assists project managers in setting realistic schedules and anticipating potential delays.

2. **Resource Allocation Optimization:**

Predictive analytics powered by AI helps optimize resource allocation by considering historical usage, project requirements, and potential bottlenecks. This ensures efficient use of manpower, equipment, and materials throughout the construction project.

3. **Cost Estimation and Budgeting:**

AI models analyze past project cost data, current market conditions, and project specifications to predict accurate cost estimates. This aids in creating realistic budgets, minimizing the risk of cost overruns during construction projects.

4. **Risk Assessment and Mitigation:**

AI-driven predictive analytics assess potential risks in construction projects by analyzing historical risk data and project-specific parameters. This enables proactive risk management and the development of mitigation strategies to address potential challenges.

5. **Material and Inventory Management:**

AI helps predict material requirements and optimize inventory levels by analyzing consumption patterns, project timelines, and supply chain data. This ensures that the right materials are available when needed, reducing delays and excess inventory costs.

6. **Equipment Health Monitoring:**

Predictive maintenance powered by AI analyzes equipment sensor data to predict when machinery or vehicles may require maintenance. This proactive approach minimizes unplanned downtime, optimizing construction project schedules.

7. **Structural Health Monitoring:**

AI is used for predictive analytics in assessing the health and integrity of structures. By analyzing sensor data, it can predict potential structural issues, enabling timely maintenance or repairs and ensuring the safety and longevity of infrastructure.

8. **Weather Impact Assessment:**

Predictive analytics, combined with AI, helps assess the potential impact of weather conditions on construction projects. By analyzing historical weather patterns and forecasts, project managers can plan activities more effectively, considering weather-related challenges.

9. **Labor Productivity Prediction:**

AI models analyze historical labor productivity data, considering factors such as working hours, conditions, and task complexity. This enables the prediction of labor productivity for future tasks, aiding in workforce planning.

10. **Supply Chain Optimization:**

AI-powered predictive analytics optimize the construction supply chain by anticipating material demands, identifying potential delays, and recommending alternative suppliers. This ensures a smooth flow of materials and minimizes disruptions.

In summary, AI-driven predictive analytics in civil engineering and construction projects enhance decision-making by providing accurate forecasts and actionable insights. From project timelines to resource allocation, cost estimation, and risk management, AI contributes to more efficient and successful construction project planning and execution.

## 4.2 Risk Management

Artificial Intelligence (AI) plays a critical role in enhancing risk management practices within the context of civil engineering and construction projects.

By leveraging advanced analytics and machine learning capabilities, AI contributes to proactive risk identification, assessment, and mitigation, particularly in the realm of project planning. Here are key aspects of AI for risk management in civil engineering and construction projects with reference to project planning:

1. **Data-driven Risk Identification:**

   AI analyzes historical project data, including past project performances, delays, and issues, to identify patterns and potential risk factors. This data-driven approach enhances the identification of both common and uncommon risks during project planning.

2. **Predictive Analytics for Risk Assessment:**

AI employs predictive analytics to assess the likelihood and impact of various risks on project timelines and budgets. By analyzing historical data, AI models can predict potential risks, allowing project managers to allocate resources and develop mitigation strategies accordingly.

3. **Dynamic Risk Registers:**

AI systems maintain dynamic risk registers that evolve as the project progresses. These registers are continuously updated based on real-time data, enabling project managers to stay informed about emerging risks and make timely decisions.

4. **Natural Language Processing (NLP) for Risk Documentation:**

AI-powered NLP tools extract insights from textual documents, such as project reports and communication logs. This aids in identifying implicit risks and ensures that all relevant risk information is considered during project planning.

5. **Scenario Analysis and Simulation:**

AI facilitates scenario analysis and simulation, allowing project managers to model different scenarios and assess the potential impact of various risks on project outcomes. This helps in developing robust risk response plans.

6. **Supplier and Vendor Risk Management:**

AI assesses the performance and reliability of suppliers and vendors by analyzing historical data and external factors. This proactive approach helps in identifying potential risks related to the supply chain and developing contingency plans.

7. **Environmental and Regulatory Risk Assessment:**

AI analyzes environmental data, regulatory changes, and compliance requirements to assess potential risks associated with the project's location and regulatory landscape. This information guides project planning to ensure adherence to standards.

### 8. Real-time Monitoring and Early Warning Systems:

AI enables real-time monitoring of project parameters, including schedule adherence, resource utilization, and external factors. Early warning systems powered by AI provide alerts for potential risks, allowing project managers to take preemptive actions.

### 9. Collaborative Risk Management:

AI fosters collaboration among project stakeholders by providing a centralized platform for sharing risk-related information. This collaborative approach ensures that insights from various perspectives contribute to comprehensive risk management strategies.

### 10. Continuous Learning and Improvement:

AI systems in risk management continuously learn from project outcomes and adjustments made during the project lifecycle. This facilitates continuous improvement in risk management practices, contributing to more effective risk mitigation in future projects.

In conclusion, AI's integration into risk management practices in civil engineering and construction projects transforms project planning by offering proactive identification, assessment, and mitigation of risks. By leveraging data-driven insights and predictive analytics, AI empowers project managers to make informed decisions and enhance the overall resilience of construction projects.

## 4.3 Scheduling and Resource Allocation

Artificial Intelligence (AI) significantly contributes to scheduling and resource allocation in civil engineering and construction projects, offering advanced tools to optimize project planning.

The integration of AI in this context enhances the efficiency, accuracy, and adaptability of scheduling and resource allocation processes. Here are key aspects of AI in scheduling and resource allocation with reference to project planning:

1. **Dynamic Scheduling Optimization:**

   AI algorithms analyze project parameters, historical data, and dependencies to dynamically optimize project schedules. This results in more realistic and adaptable schedules that account for unforeseen events and changes.

2. **Machine Learning for Predictive Scheduling:**

AI, particularly machine learning models, predicts future project timelines and identifies potential delays based on historical project data. This enables project managers to proactively address issues and adjust schedules as needed during project planning.

3. **Resource Optimization Algorithms:**

AI-powered algorithms optimize resource allocation by considering factors such as skill sets, availability, and project requirements. This ensures that the right resources are assigned to tasks, minimizing idle time and improving overall project efficiency.

4. **Predictive Analytics for Workforce Planning:**

AI analyzes historical labor data to predict workforce requirements for future tasks. This supports effective workforce planning and ensures that the right number of skilled workers is available when needed during project execution.

5. **Real-time Monitoring of Resource Usage:**

AI enables real-time monitoring of resource usage, allowing project managers to track the progress of tasks and ensure optimal resource utilization. This information helps in making data-driven decisions to enhance resource allocation.

6. **Project Constraints and Optimization:**

AI takes into account project constraints such as budget limitations, regulatory requirements, and environmental considerations when optimizing schedules and allocating resources. This ensures compliance and minimizes the risk of overruns.

7. **Collaborative Planning Platforms:**

AI fosters collaboration among project stakeholders by providing collaborative planning platforms. These platforms allow real-time sharing of project data and facilitate communication, ensuring that all team members are aligned on project scheduling and resource allocation.

8. **Scenario Analysis for Resource Contingency:**

   AI enables scenario analysis to assess the impact of potential changes on resource allocation and project schedules. Project managers can simulate different scenarios and develop contingency plans to address uncertainties.

9. **Predictive Maintenance for Equipment:**

   AI analyzes sensor data from construction equipment to predict when maintenance is needed. This proactive approach minimizes equipment downtime, ensuring that machinery is available when required in the project plan.

10. **Integration with Building Information Modeling (BIM):**

    AI integrates with BIM to enhance scheduling and resource allocation by providing a comprehensive and visual representation of the project. This facilitates better coordination and decision-making during project planning and execution.

In conclusion, AI's applications in scheduling and resource allocation revolutionize project planning in civil engineering and construction projects. By leveraging predictive analytics, optimization algorithms, and real-time monitoring, AI empowers project managers to create more resilient schedules, allocate resources efficiently, and enhance overall project success.

## 4.4 Cost Estimation

Artificial Intelligence (AI) plays a significant role in improving cost estimation processes in civil engineering and construction projects, offering advanced tools and techniques that enhance accuracy and efficiency.

When integrated into project planning, AI contributes to more precise budgeting and financial forecasting. Here are key aspects of AI for cost estimation with reference to project planning:

1. **Data-driven Cost Analysis:**

   AI leverages historical project data, including costs associated with materials, labor, equipment, and overheads, to perform data-driven cost analyses. This enables project managers to make more accurate estimates based on past project performances.

2. **Predictive Analytics for Cost Prediction:**

AI algorithms, particularly machine learning models, analyze historical cost data to predict future project costs. This predictive capability helps in anticipating potential cost overruns and adjusting project plans accordingly during the planning phase.

3. **Pattern Recognition in Cost Components:**

AI systems use pattern recognition to identify cost components within project documentation. This includes recognizing patterns in contracts, invoices, and bills of quantities, contributing to a more comprehensive understanding of project costs.

4. **Automated Quantity Takeoff:**

AI automates the process of quantity takeoff by analyzing design documents and extracting relevant information. This reduces manual effort, minimizes errors, and speeds up the estimation process during project planning.

5. **Market Conditions and Price Fluctuations:**

AI considers external factors such as market conditions and price fluctuations of construction materials. This dynamic analysis helps project managers account for potential changes in costs due to market variability.

6. **Scenario Analysis for Cost Contingency:**

AI facilitates scenario analysis to assess the impact of changes on project costs. Project managers can simulate different scenarios, evaluate the cost implications, and develop contingency plans to manage uncertainties during project planning.

7. **Risk-based Cost Estimation:**

AI integrates risk management into cost estimation by assessing the potential impact of identified risks on project costs. This proactive approach allows project managers to allocate budgets that account for uncertainties.

8. **Supplier and Vendor Performance Analysis:**

AI evaluates the performance of suppliers and vendors by analyzing historical data, ensuring that cost estimates reflect the reliability and efficiency of external partners. This contributes to more accurate budgeting during project planning.

9. **Integration with Building Information Modeling (BIM):**

AI integrates with BIM to analyze the 3D model of a construction project, extracting information relevant to cost estimation. This integration provides a more holistic understanding of project components and associated costs.

10. **Real-time Cost Monitoring:**

AI enables real-time monitoring of project costs by analyzing data from ongoing construction activities. This allows project managers to compare actual costs against estimates, identify discrepancies, and make informed decisions during project execution.

In summary, AI transforms cost estimation in civil engineering and construction projects by introducing data-driven approaches, predictive analytics, and automation. By incorporating AI into project planning, construction professionals can enhance the accuracy of cost estimates, improve budgeting processes, and mitigate financial risks for more successful project outcomes.

## 4.5 Optimization Techniques

Artificial Intelligence (AI) is a powerful tool for implementing optimization techniques in civil engineering and construction projects, particularly during project planning.

By leveraging AI-driven algorithms, these optimization techniques enhance decision-making processes, improve resource allocation, and streamline project schedules. Here are key aspects of AI for optimization in civil engineering and construction projects with reference to project planning:

1. **Resource Allocation Optimization:**

   AI algorithms analyze project requirements, resource availability, and constraints to optimize the allocation of labor, equipment, and materials. This ensures efficient resource utilization and minimizes bottlenecks during project planning.

2. **Dynamic Project Scheduling:**

AI optimizes project schedules by dynamically adjusting timelines based on real-time data, historical project performance, and dependencies. This dynamic approach enhances adaptability to changes and uncertainties during project planning.

3. **Genetic Algorithms for Project Sequencing:**

Genetic algorithms, a type of optimization algorithm inspired by natural selection, are applied to optimize project sequencing. AI-driven genetic algorithms find optimal task sequences to minimize project duration and costs.

4. **Simulated Annealing for Schedule Optimization:**

Simulated annealing, another optimization algorithm, is employed to refine project schedules by exploring different configurations and gradually moving towards optimal solutions. AI applies this technique to balance conflicting objectives and constraints.

5. **Linear Programming for Resource Optimization:**

AI utilizes linear programming techniques to optimize resource allocation by formulating and solving mathematical models. This approach considers multiple project constraints and objectives to achieve optimal resource usage.

6. **Ant Colony Optimization for Routing:**

Ant Colony Optimization algorithms are applied to optimize routing and logistics in construction projects. AI simulates the foraging behavior of ants to find the most efficient paths for transporting materials and resources.

7. **Machine Learning for Cost-Benefit Analysis:**

AI-driven machine learning models conduct cost-benefit analyses based on historical data and project parameters. This supports decision-making by identifying optimal solutions that balance costs and benefits during project planning.

8. **Constraint Programming for Project Constraints:**

AI employs constraint programming to handle complex project constraints and find optimal solutions. This technique ensures that project plans adhere to specific constraints while achieving optimization objectives.

## 9. Neural Network-based Optimization:

AI leverages neural networks to optimize complex relationships and dependencies within project planning. Neural networks can learn from historical project data to predict optimal solutions for resource allocation and scheduling.

## 10. Multi-Objective Optimization:

AI facilitates multi-objective optimization, considering multiple conflicting objectives simultaneously. This includes optimizing project schedules, resource allocation, and cost management to achieve a balanced and optimal solution.

## 11. Integration with Building Information Modeling (BIM):

AI integrates with BIM to optimize design and construction processes. BIM-based optimization considers various parameters, including material quantities, structural design, and spatial arrangements, to enhance overall project efficiency.

## 12. Real-time Monitoring and Adaptation:

AI enables real-time monitoring of project parameters and adjusts optimization strategies accordingly. This adaptive approach ensures that project plans remain optimal despite changes and uncertainties during project execution.

In conclusion, AI-driven optimization techniques in civil engineering and construction projects contribute to more effective project planning. By considering various factors, constraints, and objectives, AI helps project managers achieve optimal resource allocation, scheduling, and cost-effective solutions, ultimately enhancing the success of construction projects.

## 4.6 Real-world Examples / Use Cases

1. **Applications of AI in Project Planning:**

   a) **Project Planning Optimization (Company: ALICE Technologies):**

   - ALICE Technologies employs AI for project planning optimization. The platform uses machine learning algorithms to generate and analyze construction schedules, facilitating more efficient project planning and management.

   b) **Generative Design for Infrastructure (Company: Autodesk):**

   - Autodesk's generative design tools assist in project planning by using AI algorithms to explore and optimize design options based on specified criteria. This aids civil engineers in creating more innovative and efficient infrastructure designs.

2. **Predictive Analytics:**

   c) **Construction Project Timeline Prediction (Company: ProNovos):**

   - ProNovos utilizes predictive analytics to forecast construction project timelines. The platform analyzes historical project data and performance metrics to predict potential delays and optimize scheduling.

   d) **Equipment Performance Prediction (Company: Uptake):**

   - Uptake applies predictive analytics to monitor and predict the performance of construction equipment. The platform uses machine learning to analyze equipment data, enabling proactive maintenance and reducing downtime.

3. **Risk Management:**

   e) **Risk Identification and Mitigation (Company: Riskcast):**

   - Riskcast employs AI for risk management in construction. The platform analyzes project data to identify potential risks, assess their impact, and provides insights for effective risk mitigation strategies.

**f) Contractual Risk Assessment (Company: eCIFM Solutions - IBM TRIRIGA):**

- eCIFM Solutions, integrated with IBM TRIRIGA, utilizes AI for contractual risk assessment in construction projects. The platform applies machine learning to assess contract terms and conditions, identifying potential risks for better risk management.

## 4. Scheduling and Resource Allocation:

**g) Automated Scheduling (Company: SmartPM):**

- SmartPM employs AI for automated project scheduling in construction. The platform uses machine learning to analyze project data, optimize schedules, and allocate resources efficiently.

**h) Resource Allocation Optimization (Company: Bridgit Bench):**

- Bridgit Bench utilizes AI for resource allocation optimization. The platform analyzes historical and real-time data to provide insights into workforce capacity and helps optimize resource allocation for construction projects.

## 5. Cost Estimation:

**i) AI-Powered Cost Estimation (Company: CostOS):**

- CostOS applies AI for cost estimation in construction projects. The platform uses machine learning algorithms to analyze project data, historical costs, and market trends, providing accurate and data-driven cost estimates.

**j) Automated Quantity Takeoff (Company: STACK Construction Technologies):**

- STACK Construction Technologies employs AI for automated quantity takeoff. The platform uses machine learning to analyze project blueprints and generate accurate material quantity estimates for cost estimation.

6. **Optimization Techniques:**

k) **Project Optimization (Company: Buildots):**

- Buildots uses AI for project optimization in construction. The platform analyzes project data, progress, and performance to optimize construction processes and improve overall efficiency.

l) **Supply Chain Optimization (Company: ClearMetal):**

- ClearMetal applies AI for supply chain optimization in construction. The platform uses predictive analytics and machine learning to enhance visibility, predict delays, and optimize the construction supply chain for improved efficiency.

## 4.7 Chapter Summary: Key Points

1. AI in project planning uses predictive analytics to analyze historical data, aiding in accurate project timelines and resource estimations.

2. AI contributes to risk management by assessing potential risks based on historical data, enabling proactive identification and mitigation strategies.

3. Resource optimization is enhanced by AI algorithms, ensuring efficient allocation based on demand, availability, and project constraints.

4. In construction projects, AI aids in scheduling and sequencing by analyzing dependencies, optimizing timelines, and ensuring realistic project plans.

5. Machine learning models in project planning continuously learn from ongoing projects, adapting to changes and optimizing future planning processes.

6. AI-powered tools assist in cost estimation by analyzing historical cost data and various project parameters, reducing the likelihood of cost overruns.

7. In civil engineering, AI applications include structural analysis and design optimization, recommending efficient and cost-effective solutions.

8. AI-based document management systems organize and extract relevant information, streamlining the documentation process in project planning.

9. AI enhances collaboration by providing a centralized platform for stakeholders to access real-time project data, fostering communication and reducing errors.

10. AI applications extend to monitoring and controlling project progress, analyzing on-site sensor data and satellite imagery for real-time insights.

11. Predictive analytics in civil engineering, powered by AI, aids in project timeline prediction and efficient resource allocation.

12. AI-driven risk management involves data-driven risk identification, predictive analytics for risk assessment, and dynamic risk registers.

13. AI contributes to scheduling and resource allocation by optimizing project schedules, predicting future timelines, and ensuring efficient workforce planning.

14. AI improves cost estimation through data-driven cost analysis, predictive analytics, automated quantity takeoff, and real-time cost monitoring.

15. Optimization techniques, powered by AI, include resource allocation optimization, dynamic project scheduling, genetic algorithms, and real-time monitoring and adaptation.

## 4.8 Concept Check: Q&A Sessions

1. How does AI contribute to project planning in civil engineering construction projects?

2. What role does predictive analytics play in civil engineering and construction projects, and how is AI applied in this context?

3. How does AI enhance risk management practices in civil engineering and construction projects, particularly during project planning?

4. In what ways does AI optimize scheduling and resource allocation in civil engineering projects, and what benefits does it bring to project planning?

5. How does AI contribute to cost estimation in civil engineering and construction projects, and what key aspects does it address during project planning?

6. What optimization techniques does AI employ in civil engineering and construction projects, specifically during project planning?

7. How does AI use machine learning for cost-benefit analysis in civil engineering projects, and what benefits does it provide during project planning?

8. What role does AI play in collaborative risk management, and how does it foster communication among project stakeholders?

9. How does AI contribute to real-time monitoring and adaptation in civil engineering projects, and why is it crucial during project execution?

10. What is the significance of AI's integration with Building Information Modeling (BIM) in civil engineering and construction projects, and how does it impact project planning?

# 5. AI in Design and Modeling

Artificial Intelligence (AI) has transformed design and modeling processes in civil engineering and construction projects, offering advanced tools and techniques that streamline workflows and enhance overall efficiency.

AI applications in design and modeling span various aspects of project development.

AI facilitates generative design, allowing engineers to explore a multitude of design alternatives based on specified criteria. By analyzing large datasets and considering diverse parameters, AI-driven generative design produces innovative and optimized solutions, improving the efficiency of the design phase.

Machine learning algorithms analyze historical project data and design patterns to predict potential issues and recommend optimal design solutions. This predictive capability helps engineers identify and address challenges early in the design process, reducing the likelihood of costly revisions later.

The following block diagram illustrates the key functions of AI in Design and Modeling.

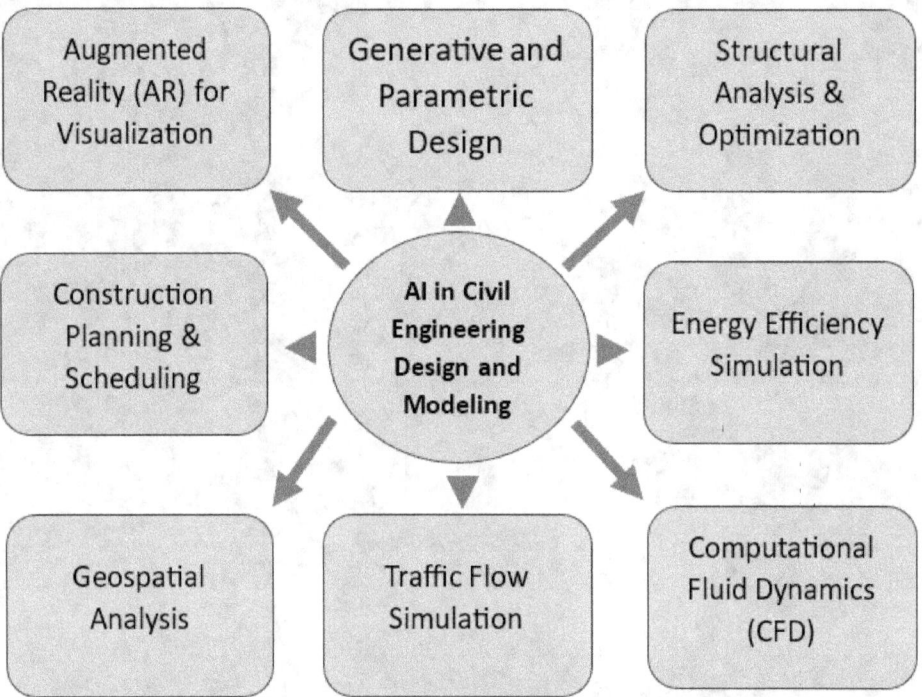

. Generative design leverages AI algorithms to rapidly explore diverse design possibilities, leading to optimized and innovative solutions. Parametric design, driven by AI, provides flexibility in design iterations, enabling quick adaptation to evolving project requirements. Structural analysis and optimization, powered by AI, enhance efficiency and performance while ensuring improved structural integrity and cost-effectiveness. Energy efficiency simulation utilizes AI to analyze and optimize the energy performance of buildings, promoting sustainable design practices. AI-driven computational fluid dynamics simulations offer an enhanced understanding of fluid-structure interactions for optimized designs. Traffic flow simulation, another AI function, optimizes transportation infrastructure designs for improved traffic management and safety. Geospatial analysis, facilitated by AI, aids in informed decision-making for site selection, land planning, and environmental considerations. AI in construction planning and scheduling optimizes schedules, predicts timelines, and ensures efficient resource utilization for improved project management. Lastly, AI-driven augmented reality applications provide a visual overlay of design models onto physical spaces, enhancing communication, understanding, and collaboration among project stakeholders. These functions collectively showcase the transformative impact of AI on design and modeling processes in civil engineering.

In structural engineering, AI contributes to structural analysis and design optimization. By automating complex calculations and simulations, AI accelerates the evaluation of structural performance, leading to more robust and cost-effective designs.

AI is employed in Building Information Modeling (BIM) systems to enhance collaboration and information exchange among project stakeholders. BIM, coupled with AI, enables real-time updates, clash detection, and comprehensive visualization, fostering a more integrated and collaborative design environment.

Natural Language Processing (NLP) enables seamless communication between engineers and AI systems. Engineers can interact with design models using natural language, making the design process more intuitive and accessible, particularly for non-technical stakeholders.

AI-driven parametric design allows engineers to create models with adjustable parameters, facilitating quick iterations and design modifications. This iterative approach supports design exploration and optimization, ensuring that the final design meets project requirements.

In geotechnical engineering, AI assists in site investigation and soil analysis. Machine learning models process geospatial data and historical information to predict soil characteristics, helping engineers make informed decisions during foundation design. In civil engineering, AI plays a pivotal role in design and modeling through various key functions.

AI contributes to traffic flow simulations and urban planning. By analyzing transportation data and considering factors such as population growth and infrastructure changes, AI models assist in designing efficient traffic systems and urban layouts.

In environmental modeling, AI aids in assessing the impact of construction projects on ecosystems. By analyzing environmental data, AI helps engineers make informed decisions that minimize the ecological footprint of construction activities.

AI-driven robotics and automation play a role in construction modeling, facilitating the use of robotic systems for tasks such as 3D printing, prefabrication, and on-site assembly. This integration enhances construction efficiency and precision.

In summary, AI in design and modeling revolutionizes civil engineering and construction projects by providing powerful tools for generative design, predictive analysis, collaborative BIM, parametric design, and more. These applications empower engineers to create innovative, optimized, and sustainable designs while improving overall project outcomes.

## 5.1 Generative Design

Generative design is a cutting-edge approach in civil engineering and construction projects that leverages artificial intelligence (AI) algorithms to explore and generate multiple design alternatives based on specified criteria and constraints.

This innovative process allows engineers to optimize and refine designs, leading to more efficient and sustainable solutions. Here are key aspects of generative design in civil engineering and construction projects:

1. **Algorithmic Exploration:**

   Generative design employs algorithms to systematically explore a vast design space. These algorithms consider a range of design parameters and constraints to generate numerous alternatives, offering a comprehensive view of possible solutions.

2. **Optimization of Design Criteria:**

The AI algorithms in generative design focus on optimizing specific design criteria, such as structural efficiency, cost-effectiveness, and environmental impact. This ensures that the generated designs align with project goals and requirements.

3. **Iterative Design Process:**

Generative design promotes an iterative design process. Engineers can review and refine generated alternatives, providing feedback to the system. This iterative cycle continues until an optimal solution meeting project objectives is identified.

4. **Complex Problem Solving:**

Generative design is particularly beneficial for addressing complex design challenges. It excels in situations where traditional manual approaches may struggle to explore the vast solution space or consider intricate relationships among design parameters.

5. **Parametric Design Exploration:**

Parametric design principles are often integrated into generative design processes. Engineers define key parameters, and the AI algorithms explore the design space by varying these parameters, allowing for a range of design possibilities.

6. **Sustainability Considerations:**

Generative design emphasizes sustainability by exploring designs that minimize environmental impact. The algorithms can optimize for factors such as energy efficiency, material usage, and waste reduction, contributing to more sustainable construction practices.

7. **Structural Optimization:**

In structural engineering, generative design is employed to optimize the layout and configuration of structural elements. The algorithms explore various arrangements to identify the most structurally efficient and cost-effective designs.

8. **Topology Optimization:**

Generative design often incorporates topology optimization, which involves determining the optimal material distribution within a given design space. This leads to lightweight and structurally efficient designs that meet performance requirements.

9. **Urban Planning and Layouts:**

In urban planning, generative design assists in exploring different city layouts and zoning configurations. The algorithms consider factors like population density, traffic flow, and green spaces to generate urban designs that balance various considerations.

10. **Collaboration and Design Exploration:**

Generative design tools facilitate collaboration among multidisciplinary teams. Engineers, architects, and other stakeholders can collectively explore design alternatives, fostering a collaborative and creative design process.

11. **Human-Centric Design:**

Generative design can incorporate human-centric factors, such as user preferences and comfort. This ensures that the final designs not only meet technical requirements but also align with the needs and preferences of end-users.

12. **Integration with Building Information Modeling (BIM):**

Generative design is often integrated with BIM systems, allowing seamless transition from generative design exploration to detailed design and construction documentation. This integration ensures continuity throughout the project lifecycle.

Generative design represents a paradigm shift in how civil engineering and construction projects approach design challenges. By harnessing the power of AI algorithms, engineers can explore innovative and optimized design solutions that may not be immediately apparent through traditional design methodologies. This approach not only enhances creativity and efficiency but also contributes to the development of more sustainable and resilient infrastructure.

## 5.2 Parametric Modeling

Parametric modeling is a powerful and versatile approach in civil engineering and construction projects that involves defining a set of parameters and rules to create and manipulate geometric shapes and structures.

This method allows for efficient design exploration, analysis, and modification by establishing relationships between different elements in a design. Here are key aspects of parametric modeling in civil engineering and construction projects:

1. **Definition of Parameters:**

   Parametric modeling begins with the definition of parameters, which are variables that control the shape, size, and other attributes of design elements. Parameters can include dimensions, angles, material properties, and more.

## 2. Relationships and Constraints:

Parametric models include relationships and constraints that govern how parameters interact with each other. These constraints ensure that changes to one parameter automatically update associated elements, maintaining design coherence.

## 3. Flexibility and Iterative Design:

Parametric modeling provides flexibility in design exploration. Engineers can iteratively adjust parameters to explore different design options quickly. This iterative process facilitates creativity and allows for the optimization of design solutions.

## 4. Efficient Design Changes:

Modifications to parametric models are efficient and systematic. If a parameter is adjusted, the model updates accordingly, ensuring that all related elements reflect the change. This efficiency is especially valuable during the design iteration phase.

## 5. Adaptability to Project Requirements:

Parametric models are adaptable to changing project requirements. As project specifications evolve, engineers can adjust parameters to meet new criteria, ensuring that the design remains aligned with project goals.

## 6. Integration with Computational Design:

Parametric modeling often integrates with computational design techniques. This involves using algorithms and scripts to automate design processes, enabling more complex and intricate geometries that respond to specific criteria.

## 7. Structural Optimization:

In structural engineering, parametric modeling is used for structural optimization. Engineers define parameters related to material properties, loads, and support conditions, allowing for the exploration of optimal structural configurations.

## 8. Generative Design Integration:

Parametric modeling is closely related to generative design. By combining the two approaches, engineers can use parametric models to establish design parameters, and then generative design algorithms can explore the design space to generate optimized alternatives.

9. **Analysis and Simulation:**

Parametric models facilitate the integration of analysis and simulation. Engineers can link parametric models with analytical tools to assess the performance of different design options under various conditions, ensuring structural integrity and compliance with standards.

10. **Urban Planning and Landscape Design:**

Parametric modeling extends to urban planning and landscape design. Designers can use parameters to control features such as building heights, street layouts, and green spaces, allowing for the exploration of diverse urban and landscape configurations.

11. **Energy and Environmental Considerations:**

Parametric models can incorporate parameters related to energy efficiency and environmental performance. Engineers can analyze how changes in design parameters affect factors like daylighting, thermal comfort, and overall sustainability.

12. **Collaborative Design:**

Parametric modeling facilitates collaborative design by allowing multiple team members to work on different aspects of the model simultaneously. Changes made by one team member can be seamlessly integrated into the overall parametric model.

In summary, parametric modeling is a versatile and efficient approach in civil engineering and construction projects. Its ability to define relationships between parameters, facilitate iterative design, and integrate with computational tools makes it a valuable methodology for creating optimized and adaptable designs.

# 5.3 Structural Analysis

Structural analysis is a critical component of civil engineering and construction projects, involving the examination and evaluation of structures to ensure their safety, stability, and compliance with design specifications.

It plays a pivotal role in the design and construction process, helping engineers understand how structures will behave under various conditions. Here are key aspects of structural analysis in civil engineering and construction projects:

1. **Load and Force Analysis:**

   Structural analysis begins with the assessment of loads and forces acting on a structure. This includes considering dead loads (permanent loads like the structure's own weight), live loads (temporary loads such as occupants or furniture), wind loads, seismic loads, and others.

2. **Equilibrium and Stability Checks:**

Engineers perform equilibrium and stability checks to ensure that a structure is in a state of static equilibrium, meaning that all forces and moments acting on the structure are balanced. Stability checks assess the structure's ability to resist overturning, sliding, and collapse.

3. **Material Properties and Behavior:**

Structural analysis involves considering the material properties of the components, such as concrete, steel, or timber. Understanding how materials behave under different loading conditions is crucial for predicting the structural response.

4. **Types of Structural Analysis:**

Different types of structural analysis methods are used, including:

- **Static Analysis:** Examines the equilibrium and stability of a structure under a set of loads.

- **Dynamic Analysis:** Investigates the dynamic response of a structure to forces like earthquakes or wind.

- **Linear and Nonlinear Analysis:** Linear analysis assumes material behavior is linear, while nonlinear analysis considers material nonlinearities.

5. **Finite Element Analysis (FEA):**

FEA is a numerical method widely used in structural analysis. It divides a complex structure into smaller elements to analyze stress, strain, and deformation. FEA is particularly valuable for assessing the behavior of intricate or irregular geometries.

6. **Structural Modeling:**

Engineers create models that represent the physical structure. These models can range from simplified 2D models for initial assessments to complex 3D models for detailed analyses. Modeling helps in visualizing and understanding the structure's behavior.

7. **Software Tools and Simulations:**

Various software tools are employed for structural analysis, allowing engineers to simulate and analyze different loading scenarios. These tools assist in performing complex calculations and visualizing the structural response.

8. **Code Compliance and Standards:**

   Structural analysis ensures that structures comply with relevant building codes and standards. Engineers must verify that the design meets safety and performance criteria set by regulatory authorities.

9. **Response to Dynamic Loads:**

   Dynamic analysis assesses how structures respond to dynamic loads, such as seismic events or wind forces. This is crucial for designing structures that can withstand environmental conditions without failure.

10. **Bridge and Infrastructure Analysis:**

    Structural analysis is extensively used in the evaluation of bridges and other infrastructure. It helps assess factors like bridge deck deflection, bending moments, and shear forces to ensure safe and efficient designs.

11. **Safety Assessments and Risk Mitigation:**

    Structural analysis includes safety assessments to identify potential failure modes and risks. Engineers can then implement risk mitigation strategies to enhance the safety and resilience of the structure.

12. **Post-Construction Monitoring:**

    In some cases, structural analysis extends to post-construction monitoring. This involves evaluating the performance of a structure in real-world conditions and addressing any unexpected issues that may arise.

In summary, structural analysis is a comprehensive process in civil engineering and construction projects, encompassing various methodologies and tools to assess the safety, stability, and performance of structures. It plays a crucial role in ensuring that buildings, bridges, and other infrastructure elements meet rigorous standards and can withstand the forces they will encounter during their lifecycle.

## 5.4 3D Printing and AI

The integration of 3D printing and artificial intelligence (AI) in civil engineering and construction projects brings about transformative changes, offering new possibilities for design, construction, and project optimization.

Here are key aspects of how 3D printing and AI intersect in the context of civil engineering and construction:

1. **Design Optimization:**

   AI algorithms can optimize designs for 3D printing by analyzing complex data sets and identifying geometries that enhance structural performance and material usage. This results in efficient and innovative designs that leverage the capabilities of 3D printing.

2. **Generative Design with AI:**

Generative design, powered by AI, can explore numerous design possibilities for structures that are suitable for 3D printing. By considering parameters and constraints, AI algorithms generate optimized designs that can be directly translated into 3D-printable forms.

3. **Customization and Parametric Design:**

AI-driven parametric design tools can create highly customized structures based on specific project requirements. When combined with 3D printing, this allows for the fabrication of unique and intricate architectural elements or components tailored to a project's needs.

4. **Material Selection and Analysis:**

AI can assist in selecting the most suitable 3D printing materials based on structural requirements, environmental conditions, and cost considerations. Additionally, AI-driven simulations can predict how printed structures will behave under different loads and conditions.

5. **Automated Printing Processes:**

AI algorithms optimize and automate the 3D printing process itself. This includes adjusting parameters in real-time, monitoring the printing environment, and making dynamic decisions to ensure the quality and integrity of the printed structure.

6. **Quality Control and Inspection:**

AI is employed for real-time quality control during 3D printing. Computer vision systems can inspect each layer as it is printed, identifying defects or deviations from the design specifications. This ensures the reliability and safety of the final structure.

7. **Construction Robotics and 3D Printing:**

AI-driven robotics are often integrated into 3D printing processes for construction. Robots equipped with AI algorithms can navigate construction sites, handle materials, and assist in the additive manufacturing of large-scale structures.

8. **Project Planning and Optimization:**

   AI assists in project planning by optimizing construction schedules, resource allocation, and budgeting for 3D printing projects. These optimization processes enhance efficiency, reduce costs, and improve overall project management.

9. **Real-time Monitoring and Adaptability:**

   AI enables real-time monitoring of 3D printing processes and adjusts parameters on the fly. This adaptability ensures that the printed structure conforms to design specifications, even in dynamic construction environments.

10. **Data-driven Decision Making:**

    AI analyzes large datasets generated during the 3D printing process and construction phases. This data-driven approach aids in decision-making, identifying patterns, predicting potential issues, and optimizing future 3D printing projects based on historical performance.

11. **Sustainability Considerations:**

    AI algorithms can be used to assess the environmental impact of 3D printing materials and processes. This allows engineers to make sustainable choices in terms of material selection and construction methods.

12. **Post-Construction Monitoring:**

    AI continues to play a role in post-construction monitoring of 3D-printed structures. Sensors and AI analytics can assess the ongoing performance and structural health, providing insights for maintenance and potential improvements.

In summary, the combination of 3D printing and AI in civil engineering and construction projects enhances the entire project lifecycle. From optimized design processes to automated construction, quality control, and ongoing monitoring, the synergy of these technologies contributes to more efficient, innovative, and sustainable construction practices.

## 5.5 Sustainable Design with AI

Sustainable design in civil engineering and construction projects involves creating structures and infrastructure that minimize environmental impact, efficiently use resources, and prioritize long-term ecological considerations.

The integration of artificial intelligence (AI) in sustainable design enhances the ability to analyze complex data, optimize designs, and make informed decisions that align with environmental goals. Here are key aspects of sustainable design with AI in civil engineering and construction projects:

1. **Energy Efficiency Optimization:**

   AI algorithms analyze building designs to optimize energy efficiency. This includes evaluating factors such as orientation, insulation, and window placement to minimize energy consumption. AI-driven simulations can assess the building's performance under various environmental conditions.

2. **Material Selection and Life Cycle Assessment:**

AI assists in selecting sustainable materials by considering environmental factors, availability, and life cycle assessments. AI algorithms evaluate the environmental impact of materials over their entire life cycle, from extraction to production, use, and disposal.

3. **Generative Design for Sustainability:**

Generative design, powered by AI, explores numerous design possibilities to identify sustainable solutions. AI algorithms consider parameters such as material efficiency, energy use, and environmental impact to generate designs that prioritize sustainability.

4. **Renewable Energy Integration:**

AI optimizes the integration of renewable energy sources, such as solar panels or wind turbines, into the design of structures. This includes determining the optimal placement and configuration of renewable energy systems to maximize their effectiveness.

5. **Waste Reduction Strategies:**

AI analyzes construction processes to identify opportunities for waste reduction. By optimizing construction methods and material usage, AI contributes to minimizing construction waste and promoting more sustainable construction practices.

6. **Water Efficiency Planning:**

AI algorithms can assess water usage patterns and optimize designs for water efficiency. This includes incorporating features like rainwater harvesting, efficient irrigation systems, and water recycling mechanisms to reduce overall water consumption.

7. **Urban Planning and Green Spaces:**

AI assists in urban planning by optimizing the layout of cities to include green spaces, pedestrian-friendly areas, and efficient transportation systems. These optimizations contribute to creating sustainable, livable urban environments.

8. **Climate Resilience Assessments:**

AI conducts climate resilience assessments to evaluate how structures will perform under changing climate conditions. This involves analyzing potential risks, such as increased temperatures, extreme weather events, or rising sea levels, and designing structures that can adapt to these challenges.

9. **Predictive Maintenance for Sustainability:**

AI-driven predictive maintenance analyzes sensor data from buildings and infrastructure to anticipate maintenance needs. This proactive approach ensures that structures remain in optimal condition, extending their lifespan and reducing the need for resource-intensive repairs or replacements.

10. **Real-time Monitoring and Adaptive Systems:**

AI enables real-time monitoring of building performance and environmental conditions. Adaptive systems can adjust building parameters, such as lighting and HVAC systems, to optimize energy use and create more sustainable operational practices.

11. **Community Engagement and Social Sustainability:**

AI can be used to analyze community needs and preferences to inform the design process. Social sustainability considerations, such as access to amenities, public spaces, and community engagement, are integrated into the design to create structures that enhance overall well-being.

12. **Regulatory Compliance and Green Building Certifications:**

AI assists in navigating regulatory requirements and achieving green building certifications. By analyzing building designs against sustainability standards, AI ensures that structures comply with environmental regulations and contribute to green building certifications.

In summary, the integration of AI in sustainable design for civil engineering and construction projects provides a comprehensive and data-driven approach to environmental considerations. From optimizing energy efficiency to reducing waste and promoting resilient urban planning, AI contributes to creating structures and infrastructure that prioritize sustainability and environmental responsibility.

## 5.6 Real-world Examples / Use Cases

1. **AI in Design and Modeling:**

    a) **Automated Design Optimization (Company: Dassault Systèmes - CATIA):**

    - Dassault Systèmes' CATIA incorporates AI for automated design optimization in civil engineering. The platform uses machine learning to enhance design processes, ensuring efficiency and precision.

    b) **AI-Enhanced Building Information Modeling (BIM) (Company: Bentley Systems):**

    - Bentley Systems employs AI in BIM for civil engineering and construction. The platform utilizes machine learning algorithms to enhance data analysis and modeling, facilitating more accurate representation of construction projects.

2. **Generative Design:**

    c) **Generative Design for Infrastructure (Company: Autodesk):**

    - Autodesk's generative design tools use AI algorithms to assist in creating innovative and optimized building designs. The technology explores various design possibilities, considering factors such as sustainability and material usage.

    d) **Topology Optimization (Company: Frustum - PTC):**

    - Frustum, now a part of PTC, applies generative design principles using AI for topology optimization. The platform explores and optimizes complex shapes and structures, improving the efficiency of design processes.

3. **Parametric Modeling:**

    e) **AI-Driven Parametric Design (Company: Kineo Computational Intelligence):**

- Kineo Computational Intelligence integrates AI into parametric design processes. The platform uses machine learning to optimize parameters, providing efficient and adaptive solutions for complex engineering designs.

f) **Parametric Design Optimization (Company: Convergent Modeling Solutions):**

- Convergent Modeling Solutions applies AI for parametric design optimization in civil engineering. The platform uses machine learning algorithms to analyze and optimize design parameters for improved performance.

4. **Structural Analysis:**

g) **AI-Powered Structural Health Monitoring (Company: COWI):**

- COWI employs AI for structural health monitoring in civil engineering. The platform uses neural networks and advanced algorithms to analyze sensor data, ensuring the continuous assessment of infrastructure stability.

h) **AI for Structural Analysis (Company: FEAmax):**

- FEAmax integrates AI into structural analysis processes. The platform utilizes machine learning algorithms to enhance the accuracy and efficiency of structural simulations, aiding in the design and evaluation of structures.

5. **3D Printing:**

i) **AI-Enhanced 3D Printing (Company: MX3D):**

- MX3D incorporates AI into 3D printing for civil engineering applications. The platform utilizes machine learning to optimize printing processes, enabling the creation of complex and customized structures.

j) **AI-Driven Concrete Printing (Company: COBOD):**

- COBOD applies AI to concrete 3D printing. The platform uses machine learning to optimize printing parameters, ensuring the construction of durable and structurally sound concrete elements.

6. **Sustainable Design:**

k) **Sustainable Building Design (Company: IESVE - Integrated Environmental Solutions):**

- IESVE integrates AI into sustainable building design processes. The platform uses machine learning to assess environmental impacts, energy efficiency, and sustainability factors, supporting the creation of eco-friendly structures.

l) **AI for Green Infrastructure Planning (Company: PlanIT Geo):**

- PlanIT Geo utilizes AI for green infrastructure planning in civil engineering. The platform applies machine learning to analyze data and optimize the planning and implementation of sustainable and green infrastructure projects.

# 5.7 Chapter Summary: Key Points

1. AI revolutionizes civil engineering design by enabling generative design, predicting challenges, and enhancing structural analysis for efficient and cost-effective solutions.

2. Generative design, powered by AI, explores diverse possibilities, optimizes criteria, and promotes iterative design processes in civil engineering projects.

3. Parametric modeling in civil engineering involves defining parameters and relationships, enabling efficient design changes, structural optimization, and collaboration among multidisciplinary teams.

4. Structural analysis is crucial for evaluating safety, stability, and compliance with standards, incorporating methods like finite element analysis and ensuring equilibrium under various loads.

5. The integration of 3D printing and AI transforms civil engineering through optimized designs, customized structures, automated printing processes, and real-time monitoring for quality control.

6. Sustainable design with AI focuses on energy efficiency, material selection, generative design for sustainability, waste reduction, and integration of renewable energy in civil engineering projects.

7. AI contributes to traffic flow simulations, urban planning, and environmental impact assessments in civil engineering, optimizing transportation systems and minimizing ecological footprints.

8. AI-driven robotics and automation enhance construction efficiency, particularly in tasks like 3D printing, prefabrication, and on-site assembly.

9. Natural Language Processing enables intuitive communication between engineers and AI systems, making design processes accessible to non-technical stakeholders.

10. In geotechnical engineering, AI assists in site investigation, soil analysis, and predicting soil characteristics for informed foundation design.

11. The collaboration of Building Information Modeling (BIM) and AI enhances real-time updates, clash detection, and comprehensive visualization for integrated and collaborative design environments.

12. Post-construction monitoring ensures ongoing structural health, and AI analytics assess performance, providing insights for maintenance and improvements.

13. AI algorithms optimize 3D printing processes, automate construction schedules, and assist in budgeting for improved project management.

14. Real-time monitoring and adaptability in 3D printing processes ensure that printed structures conform to design specifications in dynamic construction environments.

15. AI contributes to community engagement, social sustainability, and achieving green building certifications by analyzing community needs and ensuring regulatory compliance.

## 5.8 Concept Check: Q&A Session

1. What is generative design in civil engineering, and how does it leverage artificial intelligence (AI) algorithms?

2. How does parametric modeling contribute to design exploration and modification in civil engineering projects?

3. What role does structural analysis play in civil engineering, and how does it ensure the safety and stability of structures?

4. How does the integration of 3D printing and AI contribute to transformative changes in civil engineering and construction projects?

5. In sustainable design for civil engineering projects, how does AI contribute to optimizing energy efficiency?

6. What is the significance of generative design in urban planning, and how does AI-driven generative design contribute to creating sustainable urban layouts?

7. How does AI contribute to the selection of sustainable materials in civil engineering, considering factors such as environmental impact and life cycle assessments?

8. What role does AI play in the real-time monitoring and adaptability of 3D printing processes in construction projects?

9. How does AI-driven predictive maintenance contribute to sustainability in civil engineering projects?

10. What is the role of AI in assessing climate resilience in civil engineering projects, and how does it contribute to designing structures that can adapt to changing environmental conditions?

# 6. Robotics in Construction

Robotics has significantly transformed the landscape of construction, introducing automation and advanced technologies to enhance efficiency, precision, and safety in various tasks.

Automated construction equipment, such as robotic arms and excavators, now handles tasks like digging and material handling with exceptional precision and repeatability. The integration of 3D printing robots allows for the construction of complex structures layer by layer, enabling the rapid and cost-effective creation of diverse geometries, from walls to entire buildings.

In the realm of construction, robotics plays a transformative role through six key functions as shown in the following diagram.

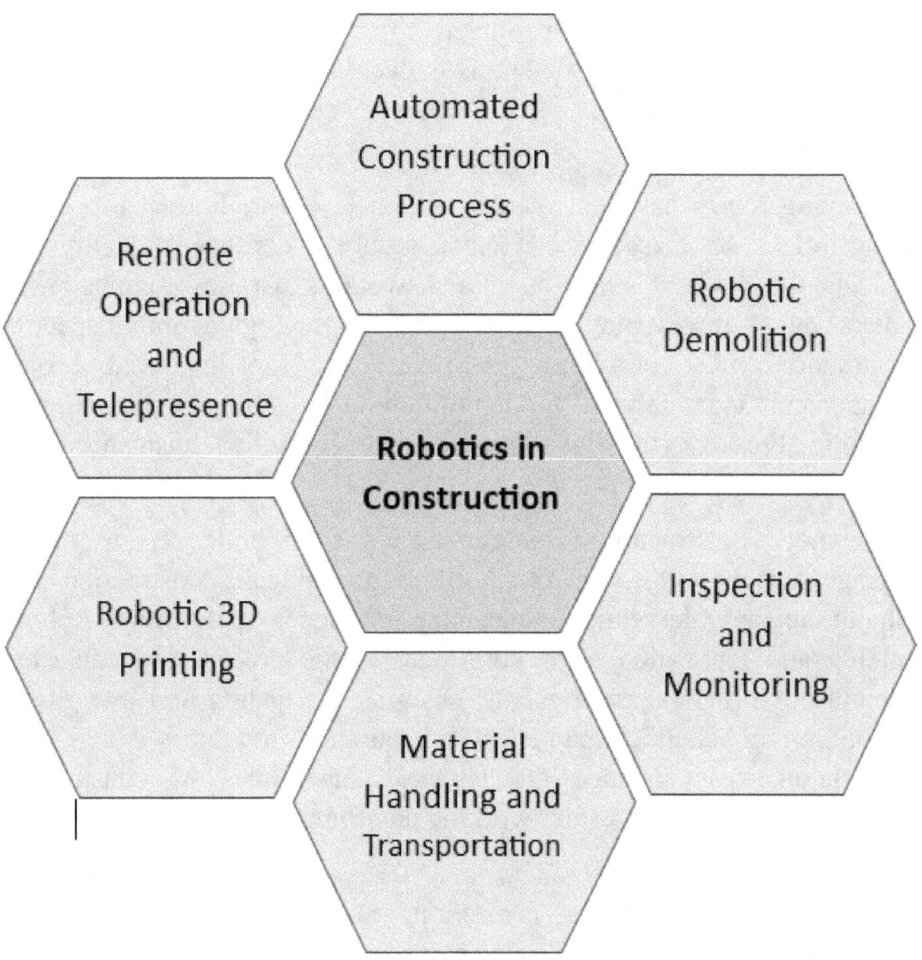

Firstly, automated construction processes employ robotic systems for tasks like bricklaying and concrete pouring, providing precision and speed while automating labor-intensive activities. Robotic demolition introduces controlled and efficient dismantling of structures, offering remote operation and enhanced safety in hazardous environments. Inspection and monitoring utilize drones and robots for site assessments, structural health monitoring, and real-time data collection, fostering informed decision-making. Material handling and transportation incorporate autonomous vehicles and robotic systems to transport and organize construction materials, reducing labor and improving efficiency. Robotic 3D printing integrates precision and customization in construction through the use of robotic arms, enabling large-scale additive manufacturing for entire structures. Lastly, remote operation and telepresence enhance accessibility and safety by employing teleoperated robotic systems in remote or hazardous environments, often coupled with virtual and augmented reality for intuitive control. These functions collectively propel the construction industry towards increased efficiency, safety, and technological advancement.

Bricklaying robots have emerged to automate the traditional process of laying bricks or masonry elements, ensuring consistent quality and reducing the physical strain on human workers. Drones equipped with cameras and sensors play a crucial role in surveying and inspecting construction sites, providing real-time data for improved project management and analysis. In demolition tasks, robots equipped with hydraulic breakers facilitate precise and controlled dismantling of structures.

Autonomous construction vehicles, ranging from bulldozers to trucks, leverage sensors, GPS, and AI algorithms to navigate construction sites without human intervention, enhancing efficiency and safety. Robotic welding and fabrication arms contribute to the precise and consistent assembly of structural components, ensuring the quality and integrity of the built environment. Human-robot collaboration introduces systems like exoskeletons that enhance the physical capabilities of construction workers, contributing to a safer working environment.

Underwater construction robots are designed for marine projects, performing tasks such as inspection, maintenance, and repair of underwater structures. Robotics also plays a pivotal role in the assembly of prefabricated building components, reducing construction time and improving overall structural quality. Automation technologies, including robots, are applied to various construction site processes, from material handling to concrete pouring, optimizing repetitive tasks and increasing overall efficiency. Finally, the implementation of remote operation and telepresence allows operators to control robots from a distance, enhancing safety in hazardous environments or challenging locations for human workers. Overall, the integration of robotics in construction represents a paradigm shift towards advanced, safer, and more efficient construction practices.

The continued evolution of robotics in construction has led to advancements in the field, shaping the industry in unprecedented ways. Construction site automation has become a common practice, with robots performing various tasks, including material handling and repetitive activities. This has not only increased efficiency but has also mitigated the risks associated with certain hazardous tasks.

The concept of the "Internet of Things" (IoT) has found its application in construction through robotic systems. These interconnected devices facilitate data sharing and communication between different elements on a construction site, enabling seamless coordination and real-time decision-making. This interconnectedness enhances overall project management and contributes to a more synchronized construction process.

In addition to automation, robots are increasingly being equipped with sensors and cameras for improved perception and decision-making capabilities. These sensors enable robots to navigate complex construction environments, avoid obstacles, and adapt to changing conditions. Such adaptive capabilities make robots versatile and suitable for a wide range of construction tasks.

The integration of Artificial Intelligence (AI) further elevates the capabilities of construction robots. AI algorithms enable robots to analyze data, learn from experiences, and make informed decisions. This intelligence is particularly valuable in tasks requiring complex problem-solving, enhancing the overall effectiveness of construction processes.

Robotic exoskeletons have gained popularity in the construction industry as wearable devices that augment the physical capabilities of human workers. These exoskeletons reduce strain and fatigue, allowing workers to perform tasks that would otherwise be physically demanding for extended periods.

In the realm of construction logistics, autonomous vehicles and drones are employed to transport materials and perform site inspections. These robotic systems optimize transportation routes, reduce delivery times, and provide valuable insights into the progress of construction projects.

Robotic systems are also contributing to the field of construction 3D printing. Large-scale 3D printers, often mounted on robotic arms, can construct entire buildings layer by layer using a variety of construction materials. This innovative approach to construction significantly reduces waste and enhances the speed of building completion.

As robotics in construction becomes more prevalent, the industry is witnessing a shift towards human-robot collaboration. This collaborative approach maximizes the strengths of both humans and robots, with robots handling repetitive or physically demanding tasks, while humans contribute their problem-solving abilities and creativity.

The emergence of swarm robotics is another notable trend. This involves the coordination of multiple robots working together collaboratively. In construction, swarm robotics can be applied to tasks such as site surveying, where a group of drones collaboratively maps and analyzes construction sites.

Looking ahead, the continued integration of robotics in construction is expected to drive innovation and efficiency. From autonomous vehicles to AI-powered robotic systems, these technologies will play a pivotal role in shaping the future of the construction industry, making processes more streamlined, sustainable, and adaptive to the evolving needs of society.

## 6.1 Autonomous Construction Vehicles

Autonomous construction vehicles represent a groundbreaking evolution in the construction industry, introducing advanced technologies that redefine traditional processes and significantly enhance project efficiency.

Here are key aspects of autonomous construction vehicles:

1. **Earthmoving and Excavation:** Autonomous bulldozers and excavators equipped with advanced sensors and navigation systems can operate without direct human control. These vehicles navigate construction sites with precision, performing tasks such as digging and material handling based on digital models.

2. **Material Transportation:** Automated trucks and haulers navigate construction sites autonomously, transporting materials with

efficiency. This minimizes the need for human drivers, enhances safety, and optimizes the overall flow of materials within the construction site.

3. **Road Construction:** In road construction, autonomous pavers and asphalt rollers operate with precision, ensuring the even and consistent application of materials. These vehicles adapt to the topography, resulting in improved road quality and reduced manual intervention.

4. **Concrete Placement:** Autonomous concrete mixers and pumps deliver precise amounts of concrete to designated locations based on digital models. This automation enhances the quality of concrete placement, reduces waste, and streamlines construction processes.

5. **Drone Surveying and Monitoring:** While not traditional vehicles, autonomous drones play a vital role in construction site surveying and monitoring. Drones equipped with cameras and sensors provide real-time data, offering insights into project progress and potential challenges.

6. **Safety Enhancement:** The deployment of autonomous construction vehicles contributes to enhanced safety on construction sites. By reducing the need for direct human operation in certain tasks, the risk of accidents and injuries associated with heavy machinery is significantly minimized.

7. **Infrastructure Maintenance:** Automated inspection vehicles equipped with sensors assess the condition of structures, helping identify maintenance needs and potential issues. This proactive approach to infrastructure monitoring contributes to the longevity and resilience of civil assets.

8. **Communication Infrastructure:** Robust vehicle-to-everything (V2X) communication systems enable autonomous vehicles to share information with each other and central control systems. This interconnectedness ensures coordinated movements, prevents collisions, and enhances overall site efficiency.

9. **Challenges and Considerations:** The widespread adoption of autonomous construction vehicles faces challenges such as regulatory frameworks, cybersecurity concerns, and the need for public

acceptance. Addressing these considerations is crucial for successful integration.

10. **Future Outlook:** As technology continues to advance, the use of autonomous construction vehicles is expected to become more prevalent, reshaping the construction industry and contributing to more efficient, safer, and technologically advanced practices. The future outlook involves overcoming challenges, refining technology, and ensuring seamless integration into various construction projects.

## 6.2 Drones in Construction

Drones in construction have emerged as versatile tools, leveraging unmanned aerial vehicle (UAV) technology to revolutionize various aspects of the construction industry.

Here are key aspects of how drones are utilized in construction:

1. **Site Surveying and Mapping:** Drones equipped with high-resolution cameras and LiDAR sensors provide efficient and accurate site surveying. They capture detailed images and data, enabling construction teams to create 3D maps and models of construction sites. This aids in precise planning and decision-making.

2. **Construction Site Inspections:** Drones are employed for regular inspections of construction sites, buildings, and structures. They can

access hard-to-reach or hazardous areas, allowing for a comprehensive assessment of construction progress, structural integrity, and adherence to safety standards.

3. **Project Documentation and Reporting:** Drones assist in project documentation by capturing visual data throughout the construction process. This documentation is valuable for progress reporting, client communication, and creating a historical record of the construction project.

4. **Topographic Mapping and Terrain Analysis:** Drones perform topographic mapping and terrain analysis, providing detailed information about the construction site's landscape. This data is crucial for accurate project planning, grading, and ensuring that construction activities align with the natural features of the land.

5. **Material Stockpile Measurement:** Construction sites often involve the management of large material stockpiles. Drones equipped with specialized software can calculate the volume of these stockpiles accurately. This technology streamlines inventory management and reduces the time required for manual measurements.

6. **Safety and Monitoring:** Drones enhance safety by monitoring construction sites for potential hazards, ensuring compliance with safety protocols, and identifying risks. Real-time monitoring allows for proactive responses to emerging safety concerns.

7. **Construction Progress Tracking:** Drones provide a bird's-eye view of construction progress over time. Construction teams can use this visual data to compare as-built conditions with project timelines, facilitating progress tracking and identifying any deviations from the original plan.

8. **Communication and Collaboration:** Drones facilitate communication and collaboration among project stakeholders. Visual data captured by drones can be shared in real-time, allowing architects, engineers, and clients to remotely view the construction site and participate in discussions.

9. **Quality Control and Defect Detection:** Drones aid in quality control by capturing high-resolution images that can be analyzed for construction defects or deviations from design specifications. Early

detection of issues allows for timely corrections, minimizing rework and associated costs.

10. **Environmental Monitoring:** Drones contribute to environmental monitoring by assessing the impact of construction activities on surrounding ecosystems. They can be used to monitor vegetation, water bodies, and other environmental factors, helping construction teams adhere to sustainability practices.

11. **Marketing and Visualization:** Drones capture stunning aerial footage and images that can be used for marketing purposes. These visuals showcase completed projects, highlight architectural features, and provide an immersive view of construction sites.

12. **Regulatory Compliance:** Drones assist in regulatory compliance by documenting construction activities and ensuring adherence to zoning and environmental regulations. This documentation can be valuable in case of audits or regulatory inspections.

In summary, drones play a multifaceted role in the construction industry, offering benefits in surveying, inspection, safety, communication, and environmental monitoring. As technology continues to advance, drones are expected to become even more integral to construction processes, contributing to improved efficiency, safety, and overall project success.

## 6.3 Robotic Process Automation in constructions

Robotic Process Automation (RPA) is a technology that uses software robots or "bots" to automate repetitive and rule-based tasks within business processes.

While RPA is more commonly associated with office-based tasks, its application in construction processes is an emerging trend that holds the potential to enhance efficiency, accuracy, and overall project success. Here are key aspects of Robotic Process Automation in construction:

1. **Automated Data Entry and Processing:**

    RPA can be employed to automate data entry tasks, reducing the need for manual input of information into various systems. This is particularly useful for processing large volumes of data related to construction projects.

2. **Document Management and Processing:**

   RPA can streamline document management by automating the sorting, organizing, and processing of construction-related documents. This includes invoices, permits, contracts, and other documentation critical to project execution.

3. **Project Scheduling and Coordination:**

   RPA tools can assist in automating project scheduling and coordination tasks. Bots can update project timelines, coordinate activities, and communicate with relevant stakeholders, ensuring that project schedules are adhered to seamlessly.

4. **Supply Chain Management:**

   RPA can optimize supply chain processes by automating the tracking of materials, managing inventory levels, and facilitating communication between suppliers and construction teams. This helps in ensuring the timely availability of materials for construction projects.

5. **Billing and Invoicing:**

   RPA can be applied to automate billing and invoicing processes. Bots can generate invoices, verify billing information, and even process payments, reducing the manual workload on administrative staff.

6. **Quality Control and Inspections:**

   RPA can contribute to quality control by automating inspection processes. Bots equipped with computer vision capabilities can identify defects or deviations from specifications during construction, ensuring that quality standards are met.

7. **Communication and Reporting:**

   RPA tools can automate communication processes by generating reports, summarizing project data, and even sending alerts or notifications based on predefined criteria. This enhances the speed and accuracy of reporting within construction projects.

8. **Health and Safety Compliance:**

RPA can assist in ensuring health and safety compliance on construction sites. Bots can automate the monitoring of safety protocols, track incidents, and generate reports to ensure that construction activities adhere to regulatory standards.

9. **Data Analytics for Decision Support:**

RPA, when integrated with data analytics tools, can provide construction managers with valuable insights for decision-making. This includes analyzing project performance, identifying trends, and predicting potential risks.

10. **Human-Machine Collaboration:**

The collaboration between human professionals and RPA tools is crucial. RPA can handle routine and repetitive tasks, allowing human workers to focus on more complex aspects of construction management, such as problem-solving and strategic decision-making.

While the application of RPA in construction is promising, it's important to address challenges such as integration with existing systems, data security, and ensuring that RPA technologies align with the specific needs and goals of construction projects. Continuous monitoring, adaptation, and strategic implementation are essential for realizing the full potential of Robotic Process Automation in the construction industry.

## 6.4 AI-assisted Construction Machinery

AI-assisted construction machinery represents a significant advancement in the construction industry, leveraging artificial intelligence technologies to enhance the efficiency, safety, and capabilities of construction equipment.

Here are key aspects of AI-assisted construction machinery:

1. **Autonomous Vehicles and Equipment:**

   AI is integrated into construction vehicles and equipment, enabling them to operate autonomously. This includes autonomous bulldozers, excavators, and other heavy machinery that can perform tasks without direct human control, improving precision and reducing manual labor.

**2. Machine Learning for Predictive Maintenance:**

AI algorithms, particularly machine learning, are used for predictive maintenance of construction machinery. By analyzing historical data, these algorithms can predict when equipment is likely to fail, allowing for proactive maintenance and minimizing downtime.

**3. Sensors and IoT Integration:**

AI-assisted construction machinery often incorporates sensors and Internet of Things (IoT) devices. These sensors collect real-time data on equipment performance, environmental conditions, and safety parameters, enabling better decision-making and monitoring on construction sites.

**4. Computer Vision for Object Recognition:**

Computer vision, a subset of AI, is employed for object recognition on construction sites. This includes the ability to identify and navigate around obstacles, detect workers in the vicinity, and ensure the safe operation of machinery.

**5. Optimized Route Planning:**

AI algorithms assist in optimizing the routes and movements of construction machinery. This is particularly valuable in large construction sites where efficient navigation can enhance productivity and reduce fuel consumption.

**6. AI in Crane Operations:**

Cranes equipped with AI technology can benefit from improved precision in lifting and placing heavy materials. AI algorithms can factor in environmental conditions and weights to optimize lifting operations, ensuring safety and efficiency.

**7. Adaptive Control Systems:**

AI-assisted machinery often features adaptive control systems. These systems can adjust the parameters of equipment operation based on real-time conditions, such as load variations, terrain changes, or unexpected obstacles.

**8. Human-Machine Collaboration:**

AI fosters collaboration between construction machinery and human operators. While certain tasks can be automated, human operators remain in control, overseeing and intervening as needed. This collaborative approach maximizes efficiency and safety.

9. **Data Analytics for Equipment Performance:**

AI facilitates data analytics for construction machinery performance. By analyzing data on equipment usage, fuel consumption, and operation patterns, construction managers can make informed decisions to optimize fleet performance.

10. **Remote Operation and Monitoring:**

AI enables remote operation and monitoring of construction machinery. Operators can control equipment from a distance, enhancing safety in hazardous environments. Additionally, real-time monitoring provides insights into machinery performance.

The integration of AI into construction machinery not only improves operational efficiency but also contributes to overall safety and sustainability in the construction industry. As technology continues to advance, AI-assisted construction machinery is expected to play a pivotal role in shaping the future of construction practices.

## 6.5 Safety Measures and Regulations

Safety measures and regulations for robotics in construction are essential to ensure the well-being of workers, prevent accidents, and promote the responsible deployment of advanced technologies.

Here are key considerations:

1. **Risk Assessment:**

   Conduct thorough risk assessments before implementing robotic systems in construction. Identify potential hazards associated with the use of robotics and assess the level of risk involved.

2. **Safety Standards Compliance:**

Adhere to established safety standards and regulations relevant to the use of robotics in construction. Ensure that robotic systems comply with industry-specific safety guidelines and standards.

3. **Worker Training:**

Provide comprehensive training programs for workers who will operate, supervise, or work alongside robotic systems. Training should cover safety protocols, emergency procedures, and the proper use of safety equipment.

4. **Emergency Stop Mechanisms:**

Implement emergency stop mechanisms on robotic equipment to allow immediate cessation of operations in case of emergencies or unforeseen situations. Ensure that these mechanisms are easily accessible to operators.

5. **Safeguarding Devices:**

Install safeguarding devices, such as barriers, fencing, or sensors, to prevent unauthorized access to robotic work areas. These devices help create a safe working environment and reduce the risk of accidents.

6. **Collaborative Robots (Cobots):**

If using collaborative robots (cobots) designed to work alongside human workers, ensure that these robots have safety features, such as force-limiting capabilities and proximity sensors, to prevent collisions and injuries.

7. **Regular Maintenance and Inspections:**

Establish a routine maintenance schedule for robotic systems to ensure they are in optimal working condition. Regular inspections help identify and address any potential safety concerns before they escalate.

8. **Safety Data and Documentation:**

Maintain documentation related to the safety features and capabilities of robotic systems. This information should be easily accessible to workers, supervisors, and safety inspectors.

9. **Coordination with Regulatory Bodies:**

Stay informed about and comply with regulations set by relevant occupational safety and health authorities. Establish communication channels with regulatory bodies to ensure that your use of robotics aligns with existing safety standards.

10. **Safety Monitoring Systems:**

Implement safety monitoring systems that provide real-time data on the status and activities of robotic systems. This allows for proactive intervention in case of anomalies or potential safety risks.

11. **Personal Protective Equipment (PPE):**

Mandate the use of appropriate personal protective equipment for workers in proximity to robotic systems. This may include safety helmets, vests, gloves, and other gear to minimize the risk of injury.

12. **Incident Reporting and Investigation:**

Establish clear procedures for reporting incidents related to robotic operations. Conduct thorough investigations into any accidents or near-misses to identify root causes and implement corrective actions.

13. **Public Safety Considerations:**

If robotic systems are deployed in areas accessible to the public, implement measures to ensure public safety. This may include signage, barriers, or restrictions to prevent unintended interactions.

14. **Continuous Improvement:**

Foster a culture of continuous improvement in safety practices related to robotics in construction. Regularly review and update safety protocols based on lessons learned and emerging best practices.

By integrating these safety measures and adhering to regulations, construction companies can harness the benefits of robotic systems while prioritizing the safety and well-being of their workforce and the broader community.

# 6.6 Real-world Examples / Use Cases

1. **Robotics in Construction:**

   A) **Bricklaying Robot (Company: Fastbrick Robotics - Hadrian X):**

   - Fastbrick Robotics developed the Hadrian X, a bricklaying robot that automates the construction of brick structures. It utilizes robotics to enhance construction speed and accuracy.

   B) **Robotic Exoskeletons (Company: Ekso Bionics):**

   - Ekso Bionics provides robotic exoskeletons for construction workers. These wearable robots enhance strength and endurance, reducing physical strain during tasks like heavy lifting and overhead work.

2. **Autonomous Construction Vehicles:**

   C) **Autonomous Excavators (Company: Built Robotics):**

   - Built Robotics retrofit standard construction equipment, such as excavators, with autonomous technology. These AI-driven vehicles operate autonomously, enhancing efficiency and safety in excavation tasks.

   D) **Autonomous Concrete Mixers (Company: Volvo Construction Equipment):**

   - Volvo Construction Equipment is exploring autonomous concrete mixers. These vehicles leverage AI and automation to optimize the mixing and pouring of concrete, improving precision and consistency.

3. **Drones in Construction:**

   E) **Construction Site Monitoring (Company: Skycatch):**

   - Skycatch utilizes drones for real-time construction site monitoring. The drones capture high-resolution images, and AI algorithms analyze the data to track progress, identify issues, and enhance project management.

   F) **Surveying and Mapping (Company: Kespry):**

- Kespry employs drones for surveying and mapping in construction. The drones equipped with AI technology capture aerial data to create accurate topographic maps and assist in site planning.

4. **Robotic Process Automation:**

   **G) Automated Document Processing (Company: Procore):**

   - Procore incorporates Robotic Process Automation (RPA) for automated document processing in construction. The platform automates routine tasks such as document routing and approval, improving efficiency.

   **H) Data Extraction and Integration (Company: ConstructConnect):**

   - ConstructConnect uses RPA for data extraction and integration in construction. The technology automates the extraction of project data from various sources and integrates it into construction management systems.

5. **AI-assisted Construction Machinery:**

   **I) AI-driven Excavators (Company: XCMG):**

   - XCMG integrates AI into excavators for construction tasks. The AI technology enhances the precision and efficiency of excavation processes, optimizing the performance of construction machinery.

   **J) AI-enhanced Bulldozers (Company: Caterpillar):**

   - Caterpillar explores AI enhancements for bulldozers in construction. The integration of AI improves the accuracy and autonomy of bulldozer operations, contributing to more efficient earthmoving tasks.

6. **Safety Measures and Regulations:**

   **K) ISO Standards for Safety (International Organization for Standardization - ISO):**

   - ISO has established standards, such as ISO 10218, for the safety of industrial robots, including those used in construction. These standards provide guidelines for the design, implementation, and operation of robotic systems to ensure safety.

## L) OSHA Regulations (Occupational Safety and Health Administration):

- OSHA in the United States regulates the use of robots in construction through safety standards. Employers must adhere to OSHA guidelines to ensure the safe deployment of robotic systems, protecting workers from potential hazards.

# 6.7 Chapter Summary: Key Points

1. Robotics in construction introduces automation, including 3D printing and bricklaying robots, enhancing precision, efficiency, and safety in various tasks.

2. Drones equipped with cameras and sensors play a crucial role in surveying and inspecting construction sites, providing real-time data for improved project management.

3. Autonomous construction vehicles, leveraging sensors and AI algorithms, navigate sites without human intervention, enhancing efficiency and safety.

4. Robotic exoskeletons reduce strain and fatigue for construction workers, augmenting physical capabilities and contributing to a safer working environment.

5. The Internet of Things (IoT) facilitates data sharing and communication between construction elements, enhancing project coordination and decision-making.

6. Robots in construction equipped with sensors and cameras navigate complex environments, adapt to changing conditions, and avoid obstacles for improved efficiency.

7. Artificial Intelligence (AI) elevates construction robots' capabilities by enabling data analysis, learning from experiences, and making informed decisions.

8. Autonomous vehicles and drones optimize construction logistics, transporting materials, performing site inspections, and providing valuable insights into project progress.

9. Construction 3D printing, with large-scale printers on robotic arms, reduces waste and speeds up building completion for more sustainable practices.

10. Human-robot collaboration maximizes strengths, with robots handling repetitive tasks and humans contributing problem-solving abilities and creativity in construction.

11. Swarm robotics coordination is applied to tasks like site surveying, where a group of drones collaboratively maps and analyzes construction sites.

12. The integration of robotics in construction contributes to a paradigm shift, driving innovation, efficiency, and safety in construction practices.

13. Autonomous construction vehicles, including bulldozers and excavators, perform tasks like digging and material handling with precision based on digital models.

14. Drones in construction revolutionize surveying, inspections, safety monitoring, and project documentation, contributing to improved efficiency and project success.

15. Robotic Process Automation (RPA) in construction automates tasks like data entry, document management, project scheduling, and supply chain management, enhancing overall efficiency and accuracy.

## 6.8 Concept Check: Q&A Session

1. How has robotics transformed the construction industry?

2. What role do drones play in construction site surveying and monitoring?

3. How do bricklaying robots contribute to the construction process?

4. In what ways do autonomous construction vehicles enhance project efficiency and safety?

5. What is the significance of the Internet of Things (IoT) in construction through robotic systems?

6. How does Robotic Process Automation (RPA) contribute to construction project success?

7. What benefits does AI bring to construction machinery?

8. How do safety measures and regulations ensure responsible deployment of robotics in construction?

9. What is the role of collaborative robots (cobots) in ensuring safety on construction sites?

10. How can construction companies balance the integration of robotic systems with public safety considerations?

# 7. AI in Structural Health Monitoring

Artificial Intelligence (AI) has emerged as a transformative tool in the field of Structural Health Monitoring (SHM) within civil engineering and construction.

SHM involves continuous monitoring and assessment of structural integrity to ensure safety, identify potential issues, and optimize maintenance strategies. AI technologies contribute significantly to enhancing the efficiency and accuracy of SHM processes.

Artificial Intelligence (AI) is revolutionizing structural health monitoring (SHM) in civil engineering and construction through eight key functions.

The following diagram illustrates the key functions of AI in Structural Health Monitoring.

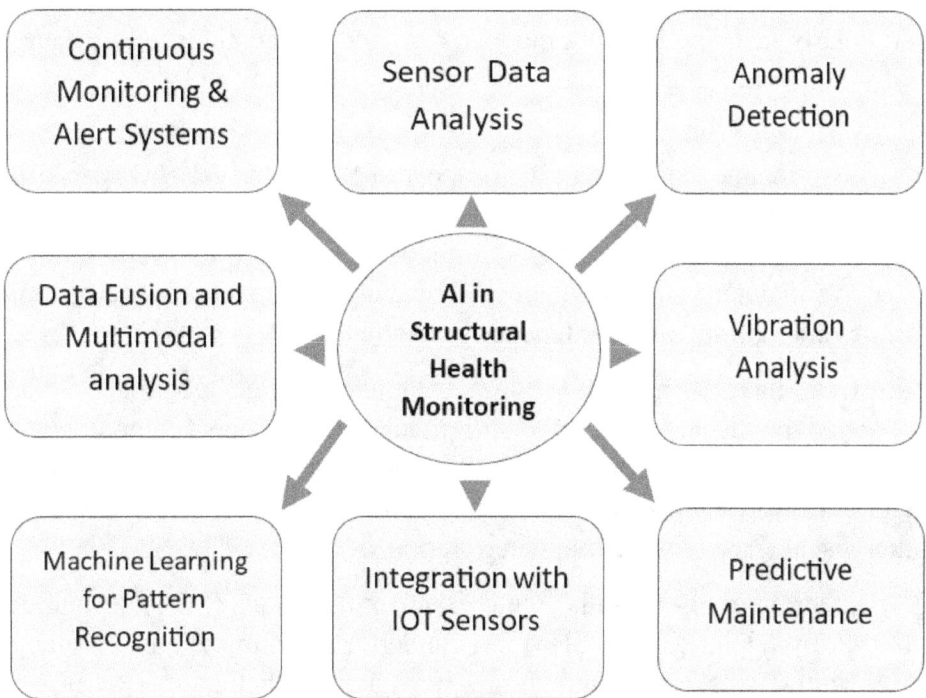

AI processes data from diverse sensors, such as accelerometers and strain gauges, providing real-time insights into structural conditions and enabling early anomaly detection. Algorithms within AI contribute to anomaly detection, identifying deviations from normal behavior to minimize the risk of structural failure through proactive maintenance. Vibration analysis using AI enhances understanding of structural dynamics and aids in stability assessment. Predictive maintenance, facilitated by AI's analysis of historical data, ensures minimized downtime and extended structure lifespan. Integration with IoT sensors allows for holistic insights by considering environmental conditions and loads. Machine learning algorithms in AI contribute to pattern recognition for more accurate diagnosis of structural issues. Data fusion and multimodal analysis by AI offer nuanced insights by combining data from various monitoring sources. Continuous monitoring and alert systems, powered by AI, enable immediate responses to potential threats, ensuring timely interventions and optimizing maintenance strategies. Collectively, these functions empower engineers to uphold structural integrity, prevent failures, and embrace a proactive and data-driven approach to SHM.

One key application of AI in SHM is the analysis of sensor data. Modern structures are equipped with various sensors, such as accelerometers and strain gauges, that generate vast amounts of data. AI algorithms, particularly machine learning models, can process and analyze this data in real-time, identifying patterns and anomalies that may indicate structural problems or changes.

Predictive maintenance is a crucial aspect of SHM, and AI plays a pivotal role in this domain. By leveraging historical data and machine learning algorithms, AI systems can predict potential structural issues, allowing for proactive maintenance interventions. This helps in preventing catastrophic failures and optimizing maintenance schedules.

AI-driven image recognition and computer vision technologies are employed in visual-based structural monitoring. Drones equipped with cameras can capture high-resolution images of structures, and AI algorithms can analyze these images to detect cracks, deformations, or other visual signs of structural deterioration.

The integration of AI with Finite Element Analysis (FEA) is another noteworthy application in SHM. AI models can optimize and accelerate FEA simulations, providing more accurate predictions of structural behavior under various conditions. This aids engineers in designing structures with improved safety and efficiency.

Sensor fusion is an advanced technique that combines data from multiple sensors to create a comprehensive understanding of structural health. AI algorithms excel in processing and interpreting fused sensor data, offering a holistic view of a structure's condition and performance.

Real-time monitoring is a critical requirement in SHM, especially for structures exposed to dynamic environmental conditions. AI algorithms enable the continuous analysis of sensor data in real-time, facilitating immediate responses to any detected abnormalities and ensuring the timely assessment of structural health.

In the context of SHM, AI is instrumental in automating the interpretation of structural health information. This includes the automatic generation of reports, identification of critical areas that require attention, and the prediction of potential future issues based on ongoing monitoring data.

Wireless Sensor Networks (WSNs) are commonly used in SHM, and AI aids in the efficient management and analysis of data from these networks. AI algorithms can optimize communication protocols, energy consumption, and data processing within WSNs, enhancing their overall effectiveness in structural monitoring.

Ethical considerations in AI-driven SHM involve ensuring the privacy and security of the data collected. As structures become more connected and data-intensive, safeguarding sensitive information and addressing potential cybersecurity threats is paramount.

In conclusion, AI is reshaping the landscape of Structural Health Monitoring in civil engineering and construction. From real-time data analysis and predictive maintenance to image recognition and sensor fusion, AI technologies contribute to more accurate, efficient, and proactive strategies for ensuring the integrity and safety of structures. The integration of AI in SHM represents a significant advancement in the field, with far-reaching implications for the design, maintenance, and sustainability of civil engineering infrastructure.

## 7.1 Sensor Technologies

Sensor technologies play a crucial role in enabling Artificial Intelligence (AI) applications for Structural Health Monitoring (SHM).

These sensors are instrumental in collecting data related to the condition, performance, and potential risks associated with structures. Here are some key sensor technologies commonly used in conjunction with AI for Structural Health Monitoring:

1. **Accelerometers:**

   - **Function:** Measure the rate of change of velocity, providing information on structural vibrations.

   - **Application:** Detect structural movements, seismic activities, and dynamic responses.

2. **Strain Gauges:**

- **Function:** Measure strain or deformation in a structure.

- **Application:** Monitor structural deformation, stress, and load distribution.

3. **Piezoelectric Sensors:**

   - **Function:** Convert mechanical stress into electrical charge.

   - **Application:** Used for monitoring structural vibrations, detecting impact, and identifying damage.

4. **Acoustic Emission Sensors:**

   - **Function:** Detect transient stress waves, often associated with the growth of cracks.

   - **Application:** Identify active damage, such as cracking or delamination, in structures.

5. **Fiber Optic Sensors:**

   - **Function:** Use variations in light to measure strain, temperature, and other parameters.

   - **Application:** Monitor strain, temperature, and deformation in critical structural elements.

6. **Load Cells:**

   - **Function:** Measure the force applied to a structure.

   - **Application:** Monitor live loads, such as traffic on bridges or buildings, and assess their impact on structural integrity.

7. **Inclinometers:**

   - **Function:** Measure the inclination or tilt of a structure.

   - **Application:** Monitor structural movements, especially in retaining walls or slopes.

8. **Temperature Sensors:**

   - **Function:** Measure temperature changes in and around a structure.

   - **Application:** Assess the impact of temperature variations on material properties.

9. **GPS and GNSS Sensors:**

- **Function:** Provide accurate location and displacement data.

- **Application:** Monitor structural movements, settlement, and deformations.

10. **Ultrasonic Sensors:**

- **Function:** Use sound waves to detect internal defects or changes in material properties.

- **Application:** Assess the integrity of materials, detect voids, and identify changes in material density.

11. **Thermographic Cameras:**

- **Function:** Capture infrared radiation to visualize temperature variations.

- **Application:** Detect thermal anomalies that may indicate structural defects or moisture ingress.

12. **Wireless Sensor Networks:**

- **Function:** Enable the deployment of a network of interconnected sensors for data collection.

- **Application:** Facilitate real-time monitoring over large areas and improve data accessibility.

Integration of AI with these sensor technologies enhances the capabilities of structural health monitoring systems. AI algorithms can analyze the vast amount of data generated by these sensors to identify patterns, anomalies, and potential risks. Machine learning models, in particular, can learn from historical data, making predictions about structural behavior and performance.

The combination of advanced sensor technologies and AI in Structural Health Monitoring contributes to more accurate and timely assessments of structural integrity, allowing for proactive maintenance and minimizing the risk of catastrophic failures.

## 7.2 Data collection and analysis

Data collection and analysis are integral components of implementing Artificial Intelligence (AI) in Structural Health Monitoring (SHM) within civil engineering and construction.

The process involves gathering relevant information from various sensors and sources, and subsequently applying AI algorithms for in-depth analysis. Here's a comprehensive overview of the data collection and analysis workflow:

1. **Data Collection:**

   a) **Sensor Deployment:**

Accelerometers, Strain Gauges, and Other Sensors: Install a network of sensors on the structure to capture data related to vibrations, deformations, and other critical parameters.

b) **Data Types:**

- **Time Series Data:** Collect continuous data over time to monitor structural changes and responses.

- **Spatial Data:** Capture data across different locations on the structure for a comprehensive assessment.

c) **Remote Sensing Technologies:**

- **Drones and UAVs:** Use aerial platforms equipped with sensors for visual inspections and data collection in hard-to-reach areas.

- **Satellite Imaging:** Employ satellite data for monitoring large-scale infrastructure and identifying potential issues.

d) **Wireless Sensor Networks:**

- **IoT Devices:** Utilize a network of interconnected sensors for real-time data transmission and collection.

- **Communication Protocols:** Implement reliable communication protocols for seamless data transfer.

e) **Environmental Sensors:**

Temperature, Humidity Sensors: Monitor environmental conditions that can impact structural health.

f) **Non-Destructive Testing (NDT):**

Ultrasonic Testing, Ground Penetrating Radar: Apply NDT methods for assessing internal structural conditions without causing damage.

**Data Analysis:**

a) **Preprocessing:**

- **Noise Removal:** Eliminate irrelevant data or noise from sensor readings.

- **Normalization:** Standardize data to ensure consistency in scale and units.

b) **Feature Extraction:**

- **Identify Relevant Features:** Select key parameters from the dataset for analysis.

- **Transformations:** Apply mathematical transformations to enhance feature representation.

c) **Machine Learning Models:**

- **Supervised Learning:** Train models on labeled datasets to predict specific outcomes (e.g., structural damage).

- **Unsupervised Learning:** Discover patterns and anomalies in data without predefined labels.

d) **Deep Learning:**

- **Neural Networks:** Utilize deep learning architectures for complex pattern recognition.

- **Convolutional Neural Networks (CNNs):** Apply CNNs for image-based structural assessments.

e) **Time Series Analysis:**

- **Temporal Patterns:** Analyze trends and patterns over time to predict structural behavior.

- **Frequency Analysis:** Assess the frequency components of vibrations for anomaly detection.

f) **Statistical Methods:**

- **Descriptive Statistics:** Summarize and describe the main features of the data.

- **Regression Analysis:** Explore relationships between variables and predict future trends.

g) **Anomaly Detection:**

- **Threshold-Based Methods:** Set thresholds for normal behavior and identify anomalies.

- **Machine Learning-Based Anomaly Detection:** Train models to recognize unusual patterns.

h) **Predictive Maintenance:**

- **Remaining Useful Life (RUL) Prediction:** Forecast the remaining lifespan of structural components.

- **Failure Probability Analysis:** Estimate the likelihood of future failures.

i) **Visualization:**

- **3D Models and Graphs:** Visualize structural conditions using graphical representations.

- **Dashboard Interfaces:** Develop user-friendly interfaces for real-time monitoring.

j) **Integration with Decision Support Systems:**

- **Automated Alerts:** Implement automated alert systems for immediate response to critical structural changes.

- **Integration with BIM (Building Information Modeling):** Enhance collaboration and decision-making with BIM-integrated data.

The iterative process of data collection and analysis, coupled with continuous learning from AI models, allows for a proactive approach to structural health management in civil engineering and construction. It facilitates early detection of issues, predictive maintenance, and improved decision-making for ensuring the longevity and safety of infrastructure.

## 7.3 Early Warning Systems

Implementing Early Warning Systems (EWS) is a crucial aspect of applying Artificial Intelligence (AI) in Structural Health Monitoring (SHM) within civil engineering and construction.

Early Warning Systems aim to provide timely alerts and insights into potential structural issues, enabling proactive intervention. Here's an overview of the components and considerations for developing EWS using AI:

1.  **Components of Early Warning Systems:**

    i.    **Sensor Network:**

Deploy a network of sensors, including accelerometers, strain gauges, and environmental sensors, to collect real-time data on structural behavior.

ii.   **Data Acquisition System:**

Implement a robust system for collecting, storing, and transmitting data from sensors to a centralized platform.

iii.   **Communication Infrastructure:**

Establish a reliable communication network, potentially using IoT devices, to facilitate seamless data transmission from sensors to the monitoring system.

iv.   **Data Preprocessing:**

Apply preprocessing techniques to clean and normalize data, removing noise and ensuring consistency.

v.   **Feature Extraction:**

Identify relevant features from sensor data that are indicative of structural health conditions.

vi.   **Machine Learning Models:**

Utilize machine learning algorithms, such as supervised learning models or deep learning architectures, to train on historical data and predict structural behavior.

vii.   **Anomaly Detection:**

Employ anomaly detection algorithms to identify deviations from expected structural behavior.

viii.   **Thresholds and Criteria:**

Set threshold values and criteria for defining normal and abnormal structural conditions.

ix.   **Predictive Analytics:**

Implement predictive analytics models to forecast potential issues, including Remaining Useful Life (RUL) predictions.

x.   **Decision Support Systems:**

Integrate with decision support systems to provide actionable insights based on AI predictions.

## 2. Considerations for Developing Early Warning Systems:

### i. Real-Time Monitoring:

Ensure that the Early Warning System provides real-time monitoring capabilities to detect structural changes promptly.

### ii. Scalability:

Design the system to be scalable, allowing for the integration of additional sensors or the expansion of monitored areas.

### iii. Adaptability:

Develop AI models that can adapt to changes in environmental conditions, loads, and other factors influencing structural health.

### iv. Human-Machine Collaboration:

Enable collaboration between AI algorithms and human experts, allowing for a combined approach to decision-making.

### v. Automated Alerts:

Implement automated alert systems that trigger notifications when anomalies or potential issues are detected.

### vi. Integration with Building Information Modeling (BIM):

Integrate EWS with BIM for a holistic view of the structure and improved decision-making.

### vii. Historical Data Analysis:

Leverage historical data to train and continuously improve AI models, enhancing their accuracy and reliability.

### viii. Redundancy and Reliability:

Build redundancy into the system to ensure reliability even in the case of sensor failures or communication issues.

ix. **Regulatory Compliance:**

Ensure that the Early Warning System complies with relevant safety and regulatory standards in the construction industry.

x. **User-Friendly Interface:**

Develop a user-friendly interface for stakeholders, including engineers, project managers, and maintenance personnel, to easily interpret and act upon the information provided by the EWS.

xi. **Cybersecurity Measures:**

Implement robust cybersecurity measures to protect the integrity and confidentiality of the data collected and transmitted by the Early Warning System.

xii. **Continuous Improvement:**

Establish mechanisms for continuous improvement, incorporating feedback from system users and updating AI models as needed.

By integrating these components and considerations, an Early Warning System powered by AI in Structural Health Monitoring can play a pivotal role in enhancing the safety, resilience, and longevity of civil engineering structures. It enables a proactive approach to maintenance and risk mitigation, minimizing the impact of potential structural issues on infrastructure projects.

## 7.4 Predictive Maintenance

Predictive maintenance, enabled by Artificial Intelligence (AI) in Structural Health Monitoring (SHM) within civil engineering and construction, is a proactive strategy to anticipate and address potential issues before they lead to structural failures.

Here's an overview of the key aspects of predictive maintenance using AI:

1.  **Components of Predictive Maintenance with AI in SHM:**

    i.  **Sensor Deployment:**

        Install a network of sensors to collect real-time data on structural conditions. These sensors can include accelerometers, strain gauges, temperature sensors, and other relevant monitoring devices.

ii. **Data Collection and Storage:**

Implement a robust data collection and storage system to capture sensor data continuously. Historical data is crucial for training predictive models.

iii. **Data Preprocessing:**

Apply preprocessing techniques to clean, normalize, and organize the collected data, ensuring its quality and reliability.

iv. **Feature Extraction:**

Identify key features from the sensor data that are indicative of structural health conditions. Relevant features may include vibration patterns, strain levels, and environmental factors.

v. **Machine Learning Models:**

Utilize machine learning algorithms, such as regression models, decision trees, or more advanced techniques like neural networks, to train on historical data and predict future structural behavior.

vi. **Failure Mode Analysis:**

Identify potential failure modes and associated patterns in historical data to inform the development of predictive models.

vii. **Anomaly Detection:**

Implement anomaly detection algorithms to identify deviations from normal structural behavior, which may indicate the early stages of deterioration.

viii. **Predictive Analytics:**

Develop models that provide predictions related to structural performance, remaining useful life (RUL), and the likelihood of specific failure modes occurring.

ix. **Decision Support Systems:**

Integrate with decision support systems that use AI predictions to guide maintenance planning and intervention strategies.

x. **Risk Assessment:**

Conduct risk assessments based on predictive models to prioritize maintenance activities for structures with higher likelihoods of issues.

2. **Benefits of Predictive Maintenance with AI in SHM:**

i. **Cost Savings:**

Proactively addressing maintenance needs based on predictive insights can lead to cost savings compared to reactive maintenance after failures occur.

ii. **Increased Safety:**

By identifying potential structural issues early, predictive maintenance contributes to increased safety by minimizing the risk of unexpected failures.

iii. **Extended Asset Lifespan:**

Optimizing maintenance schedules based on predictive analytics can contribute to the extended lifespan of civil engineering structures.

iv. **Minimized Downtime:**

Predictive maintenance allows for planned and scheduled interventions, minimizing downtime associated with emergency repairs.

v. **Resource Efficiency:**

Resources such as labor and materials can be used more efficiently, focusing efforts on structures that require immediate attention.

vi. **Data-Driven Decision-Making:**

Decision-making processes are informed by data-driven insights, enhancing the precision and accuracy of maintenance planning.

### vii. Improved Infrastructure Resilience:

Proactive maintenance measures contribute to overall infrastructure resilience, ensuring structures can withstand environmental and operational challenges.

### viii. Adaptive Maintenance Strategies:

Predictive maintenance models can adapt to changing conditions, allowing for dynamic adjustments to maintenance strategies based on evolving structural health data.

## 3. Considerations for Implementing Predictive Maintenance with AI in SHM:

### i. Data Quality:

Ensure the quality and reliability of the data collected, as predictive models heavily depend on accurate and representative information.

### ii. Continuous Monitoring:

Implement continuous monitoring to capture real-time changes in structural conditions and update predictive models accordingly.

### iii. Integration with Maintenance Workflows:

Integrate predictive maintenance strategies seamlessly into existing maintenance workflows to ensure practical implementation.

### iv. Human Expertise:

Combine AI insights with human expertise to make informed decisions and interpret complex structural health data.

### v. Regulatory Compliance:

Ensure that predictive maintenance strategies comply with relevant safety and regulatory standards in the construction industry.

### vi. Cybersecurity Measures:

Implement robust cybersecurity measures to protect the integrity and confidentiality of the data used for predictive maintenance.

Predictive maintenance, fueled by AI in SHM, represents a forward-thinking approach to infrastructure management, aligning with the industry's goals of safety, efficiency, and sustainability.

## 7.5 Challenges of AI in SHM

Implementing Artificial Intelligence (AI) in Structural Health Monitoring (SHM) presents several challenges that need to be addressed for successful integration and optimal performance.

Here are key challenges associated with AI in SHM within civil engineering and construction:

1. **Data Quality and Quantity:**

   - **Challenge:** Insufficient or poor-quality data can hinder the effectiveness of AI models. In some cases, there may be limited historical data for training robust machine learning algorithms.

- **Mitigation:** Implement rigorous data collection processes, ensure data integrity, and explore strategies for augmenting datasets, such as simulation or synthetic data generation.

2. **Model Complexity:**

- **Challenge:** Developing accurate AI models for SHM can be complex due to the intricate nature of structural behaviors and the variety of factors influencing them.

- **Mitigation:** Balance model complexity with interpretability, and employ model validation techniques. Collaborate with domain experts to refine models and enhance their accuracy.

3. **Interpretability and Explainability:**

- **Challenge:** AI models, particularly deep learning models, can be perceived as "black boxes" with limited interpretability. Understanding model decisions is crucial, especially in critical applications like structural monitoring.

- **Mitigation:** Prioritize the use of interpretable models, incorporate explainability techniques, and engage in interdisciplinary collaborations to enhance model understanding.

4. **Sensor Selection and Placement:**

- **Challenge:** Selecting appropriate sensors and determining their optimal placement is challenging. Inadequate sensor coverage or improper placement can lead to incomplete or inaccurate data.

- **Mitigation:** Conduct thorough sensor placement studies, consider redundancy, and leverage multiple types of sensors to capture a comprehensive view of structural health.

5. **Data Privacy and Security:**

- **Challenge:** Handling sensitive structural health data raises concerns about privacy and security. Unauthorized access or data breaches could compromise the integrity of the monitoring system.

- **Mitigation:** Implement robust cybersecurity measures, encrypt data during transmission and storage, and adhere to data privacy regulations and standards.

6. **Computational Resources:**

   - **Challenge:** AI models, especially complex ones, may require significant computational resources. Deploying resource-intensive models on-site can be challenging, particularly for edge devices.

   - **Mitigation:** Optimize models for efficiency, explore edge computing solutions, and balance the trade-off between model accuracy and computational requirements.

7. **Real-time Processing:**

   - **Challenge:** Some applications, such as real-time monitoring of dynamic structures, demand extremely low-latency processing. Achieving real-time capabilities can be challenging, especially with large datasets.

   - **Mitigation:** Optimize algorithms for real-time performance, leverage parallel processing, and consider edge computing solutions to reduce latency.

8. **Adaptability to Environmental Changes:**

   - **Challenge:** Environmental conditions, such as temperature variations and external loads, can impact structural health. AI models must adapt to changing conditions for accurate predictions.

   - **Mitigation:** Develop models that account for environmental variables, regularly update models based on changing conditions, and employ adaptive learning techniques.

9. **Regulatory Compliance:**

   - **Challenge:** Compliance with industry standards and regulations, such as safety standards and data protection laws, is essential. Lack of alignment with these standards can hinder widespread adoption.

- **Mitigation:** Stay informed about relevant regulations, involve legal experts in the implementation process, and ensure that AI applications comply with established standards.

### 10. Cost and Accessibility:

- **Challenge:** The initial costs associated with implementing AI in SHM, including sensor installation and AI model development, can be a barrier, especially for smaller projects.

- **Mitigation:** Demonstrate the long-term benefits and return on investment, explore cost-sharing models, and foster collaboration between stakeholders to make AI in SHM more accessible.

Addressing these challenges requires a multidisciplinary approach, involving collaboration between AI experts, structural engineers, data scientists, and regulatory specialists. Continuous research, development, and industry-wide initiatives are essential for overcoming these challenges and unlocking the full potential of AI in Structural Health Monitoring.

# 7.6 Real-world Examples / Use Cases

1. **AI in Structural Health Monitoring:**

   **A) Automated Structural Assessment (Company: DeepSig):**

   - DeepSig employs AI for automated structural health monitoring. The platform uses machine learning algorithms to analyze data from sensors, providing real-time insights into the condition of civil infrastructure.

   **B) AI-driven Crack Detection (Company: Respeecher):**

   - Respeecher applies AI for crack detection in civil infrastructure. The platform utilizes machine learning to analyze sensor data and identify potential structural issues, enabling proactive maintenance.

2. **Sensor Technologies:**

   **C) Fiber Optic Sensors (Company: Luna Innovations):**

   - Luna Innovations provides fiber optic sensors for structural health monitoring. These sensors use light-based technology to measure strain, temperature, and other structural parameters, enhancing the accuracy of monitoring systems.

   **D) Wireless Sensor Networks (Company: Acellent Technologies):**

   - Acellent Technologies offers wireless sensor networks for structural health monitoring. The sensors are embedded in structures to collect data on vibrations, stress, and other factors, facilitating real-time monitoring.

3. **Data Collection and Analysis:**

   **E) Smart Infrastructure Platform (Company: CEMEX):**

   - CEMEX's smart infrastructure platform utilizes data collection and analysis for structural health monitoring. The platform integrates sensors and AI to gather data on concrete strength, temperature, and other factors, enhancing construction quality.

**F) Cloud-based Structural Monitoring (Company: Senceive):**

- Senceive provides cloud-based solutions for structural health monitoring. The platform collects data from wireless sensors, and AI algorithms analyze the information to assess the structural condition of assets.

4. **Early Warning Systems:**

**G) Acoustic Emission Monitoring (Company: MISTRAS Group):**

- MISTRAS Group utilizes acoustic emission monitoring for early warning in structural health monitoring. The technology detects the release of stress waves, providing early indications of potential structural issues.

**H) Vibration-based Early Warning (Company: Kinemetrics):**

- Kinemetrics offers vibration-based early warning systems for structural health monitoring. The platform uses accelerometers and seismometers to detect vibrations and assess structural integrity, enabling timely warnings.

5. **Predictive Maintenance:**

**I) Predictive Maintenance Analytics (Company: Senseye):**

- Senseye applies AI for predictive maintenance in structural health monitoring. The platform analyzes sensor data to predict potential equipment failures and optimize maintenance schedules, ensuring the longevity of civil infrastructure.

**J) AI-driven Bridge Maintenance (Company: IBM Watson Health):**

- IBM Watson Health employs AI for predictive maintenance of bridges. The platform analyzes data from sensors and inspections to predict maintenance needs, enhancing the safety and reliability of bridge infrastructure.

## 7.7 Chapter Summary: Key Points

1. AI in Structural Health Monitoring (SHM) optimizes maintenance, ensuring safety and early anomaly detection through real-time analysis of sensor data.

2. Machine learning models enhance vibration analysis, contributing to stability assessments and a proactive approach to structural maintenance.

3. Predictive maintenance, a critical SHM aspect, leverages AI to predict potential structural issues, preventing catastrophic failures and extending structure lifespan.

4. AI-driven image recognition and computer vision aid in visual-based structural monitoring, detecting defects using drones and cameras.

5. Integration of AI with Finite Element Analysis (FEA) accelerates simulations, improving predictions of structural behavior and safety.

6. Sensor fusion combines data from various sensors, enabling nuanced insights into a structure's condition and performance through AI algorithms.

7. Real-time monitoring by AI algorithms facilitates immediate responses to abnormalities, optimizing maintenance strategies for timely interventions.

8. Automation of structural health information interpretation by AI includes report generation, identification of critical areas, and prediction of future issues.

9. Wireless Sensor Networks (WSNs) benefit from AI optimization in managing and analyzing data, enhancing structural monitoring effectiveness.

10. Ethical considerations in AI-driven SHM involve safeguarding data privacy and addressing cybersecurity threats associated with connected structures.

11. Accelerometers measure velocity changes, providing insights into structural vibrations, seismic activities, and dynamic responses.

12. Piezoelectric sensors convert mechanical stress to electrical charge, monitoring vibrations, impact, and structural damage.

13. Acoustic emission sensors detect stress waves, identifying active damage like cracking or delamination in structures.

14. Fiber optic sensors use light variations to measure strain, temperature, and deformation in critical structural elements.

15. Predictive maintenance with AI in SHM offers cost savings, increased safety, extended asset lifespan, and adaptive maintenance strategies, aligning with industry goals.

## 7.8 Concept Check: Q&A Session

1. What is the primary purpose of Structural Health Monitoring (SHM) in civil engineering and construction?

2. How does Artificial Intelligence (AI) contribute to anomaly detection in SHM, particularly in analyzing sensor data?

3. What is the significance of predictive maintenance in the context of SHM, and how does AI play a role in this aspect?

4. How do drones equipped with AI-driven image recognition contribute to visual-based structural monitoring?

5. What is the role of AI in the integration with Finite Element Analysis (FEA) in SHM?

6. How do AI-driven Wireless Sensor Networks (WSNs) contribute to efficient management and analysis of data in SHM?

7. What are the ethical considerations associated with AI-driven SHM, particularly concerning data privacy and security?

8. How do AI models assist in automating the interpretation of structural health information in SHM?

9. What are the key sensor technologies commonly used in conjunction with AI for SHM, and how do they contribute to data collection?

10. What challenges are associated with implementing AI in SHM, and how can these challenges be addressed?

# 8. Smart Infrastructure and IoT

Smart infrastructure, coupled with the Internet of Things (IoT), is revolutionizing civil engineering and construction practices, ushering in an era of interconnected, data-driven solutions.

In this paradigm, physical infrastructure is embedded with sensors, actuators, and communication technologies that enable real-time monitoring, analysis, and control.

The integration of IoT in civil engineering allows for the creation of Smart Cities, where various elements of urban infrastructure are interconnected to enhance efficiency and sustainability. From smart transportation systems to intelligent energy grids, IoT in civil engineering contributes to more optimized and resilient urban environments.

Structural health monitoring is a key application of IoT in civil engineering. Sensors embedded in structures, such as bridges and buildings, provide continuous data on factors like strain, temperature, and vibrations. This real-time information allows engineers to assess structural integrity, identify potential issues, and implement preventive maintenance strategies.

Smart transportation systems leverage IoT technologies to enhance efficiency and safety. Connected vehicles, traffic signals, and infrastructure can communicate with each other, optimizing traffic flow, reducing congestion, and improving overall transportation planning and management.

IoT-driven construction sites utilize sensors and wearable devices to monitor worker safety, equipment performance, and project progress. Real-time data from these sources allows for immediate responses to potential safety hazards, efficient resource allocation, and improved project timelines.

Environmental monitoring is another domain where IoT plays a crucial role in civil engineering. Sensors can measure air quality, noise levels, and water quality, providing valuable data for managing environmental impacts and ensuring compliance with regulatory standards.

Asset management in civil engineering benefits from IoT by enabling the tracking and monitoring of infrastructure components. Smart sensors on assets such as pipelines, roads, and utility networks help assess their condition, predict maintenance needs, and optimize the lifecycle management of these assets.

Energy efficiency is a focus area in smart infrastructure. IoT-enabled systems allow for the monitoring and control of energy consumption in buildings, street lighting, and other infrastructure elements, contributing to sustainability goals and cost savings.

Emergency response and disaster management benefit from IoT in civil engineering. Connected sensors and monitoring systems provide early warnings for potential disasters, enabling swift responses and minimizing the impact on infrastructure and public safety.

IoT platforms and data analytics are essential components of smart infrastructure. The collected data from various sensors are processed and analyzed to derive actionable insights. These insights inform decision-making, allowing for more informed planning, efficient resource allocation, and proactive maintenance.

Smart infrastructure and the Internet of Things (IoT) are reshaping the landscape of civil engineering and construction with the following eight key functions.

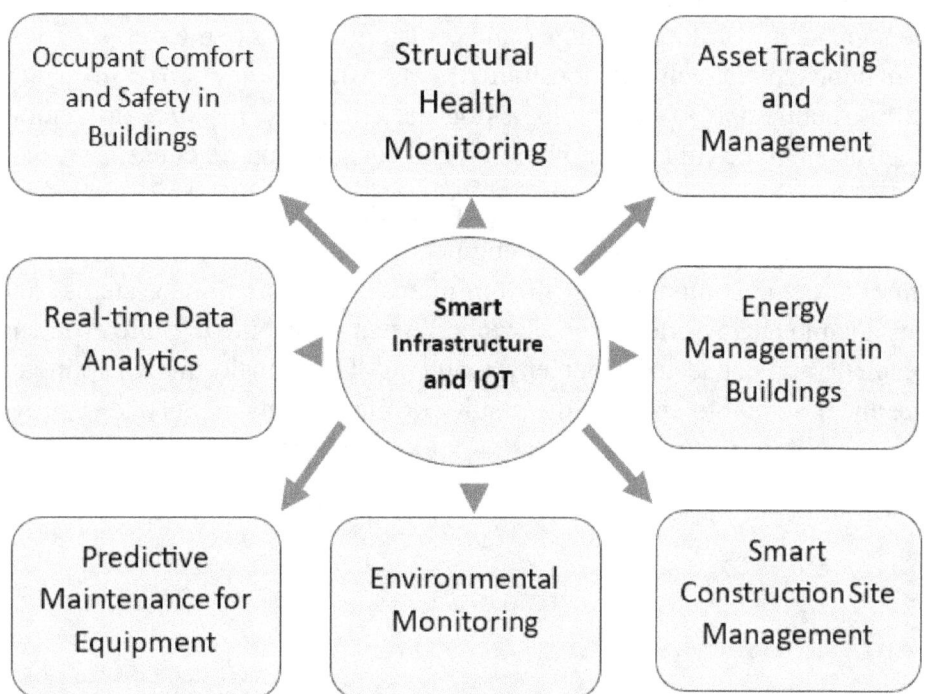

Structural health monitoring involves the integration of sensors into structures, enabling real-time assessment for early issue detection, proactive maintenance, and heightened safety. Asset tracking and management utilize IoT devices to monitor and manage construction equipment and materials, offering enhanced visibility, reduced theft, and optimized resource utilization. Energy management in buildings employs IoT sensors to monitor and optimize energy usage, resulting in increased efficiency, cost reduction, and compliance with sustainability standards. Smart construction site management deploys IoT devices for real-time monitoring, leading to improved project management, enhanced safety, and efficient resource allocation. Environmental monitoring utilizes sensors to track conditions on construction sites, ensuring compliance with regulations, proactive risk management, and sustainable practices.

Predictive maintenance for equipment integrates IoT sensors to monitor and predict maintenance needs, minimizing downtime and optimizing equipment lifespan. Real-time data analytics involves continuous data collection and analysis from various sensors, facilitating informed decision-making, early issue identification, and process optimization. Finally, occupant comfort and safety in buildings are improved through the implementation of IoT devices, enhancing building design, providing personalized comfort settings, and bolstering safety measures. These functions collectively demonstrate the transformative impact of smart infrastructure and IoT technologies on the efficiency, resource utilization, and sustainability of civil engineering and construction practices.

In conclusion, the convergence of smart infrastructure and IoT is reshaping the landscape of civil engineering and construction. From real-time structural monitoring to optimized transportation systems and sustainable energy practices, IoT technologies contribute to more efficient, resilient, and connected urban environments. This transformative approach has the potential to enhance the quality of life for inhabitants and improve the overall sustainability of our cities and infrastructure.

# 8.1 Internet of Things (IoT)

The Internet of Things (IoT) plays a pivotal role in transforming civil engineering and construction practices, leading to the development of smart infrastructure.

Here are key aspects of utilizing IoT for smart infrastructure in civil engineering and construction:

1. **Sensors and Monitoring:**

   - **Application:** Embedding sensors in infrastructure elements (e.g., bridges, buildings, pipelines) for real-time monitoring of structural health, environmental conditions, and performance.

   - **Benefits:** Early detection of potential issues, predictive maintenance, and improved safety through continuous monitoring.

2. **Asset Tracking and Management:**

   - **Application:** Using RFID or GPS-enabled devices to track and manage construction equipment, materials, and components.

   - **Benefits:** Enhanced visibility into the location and status of assets, optimized logistics, and reduced losses.

3. **Energy Management:**

   - **Application:** Implementing IoT-enabled systems to monitor and control energy usage in buildings and infrastructure.

   - **Benefits:** Improved energy efficiency, cost savings, and sustainability through smart energy management.

4. **Smart Buildings:**

   - **Application:** Deploying IoT devices for building automation, including lighting, HVAC systems, and security.

   - **Benefits:** Energy conservation, occupant comfort, and optimized building operations through intelligent control systems.

5. **Environmental Monitoring:**

   - **Application:** Utilizing IoT sensors to monitor air and water quality, noise levels, and other environmental factors.

   - **Benefits:** Early identification of environmental impacts, compliance with regulations, and sustainable construction practices.

6. **Construction Site Safety:**

   - **Application:** Implementing wearable devices and sensors to monitor workers' health and safety on construction sites.

   - **Benefits:** Improved safety outcomes, real-time incident detection, and data-driven safety protocols.

7. **Supply Chain Optimization:**

   - **Application:** Integrating IoT in the supply chain to track materials, monitor inventory levels, and optimize procurement processes.

- **Benefits:** Reduced delays, minimized waste, and enhanced efficiency in the construction supply chain.

8. **Traffic Management:**

   - **Application:** Deploying IoT devices, such as smart traffic lights and sensors, for intelligent traffic management.

   - **Benefits:** Improved traffic flow, reduced congestion, and enhanced transportation infrastructure efficiency.

9. **Remote Monitoring and Control:**

   - **Application**: Enabling remote monitoring and control of infrastructure systems through IoT-connected devices.

   - **Benefits:** Increased operational flexibility, rapid response to issues, and reduced downtime.

10. **Data Analytics and Decision Support:**

    - **Application:** Leveraging data collected from IoT devices for advanced analytics to inform decision-making.

    - **Benefits:** Informed project planning, risk mitigation, and optimized performance based on actionable insights.

11. **Smart Water Management:**

    - **Application:** Using IoT sensors to monitor water infrastructure, detect leaks, and optimize water distribution.

    - **Benefits:** Conservation of water resources, early leak detection, and improved water infrastructure resilience.

12. **Predictive Maintenance:**

    - **Application:** Implementing IoT-based predictive maintenance systems to monitor equipment health and schedule maintenance proactively.

    - **Benefits:** Reduced downtime, extended equipment lifespan, and cost-effective maintenance practices.

13. **Integration with BIM (Building Information Modeling):**

    - **Application:** Integrating IoT data with BIM models for comprehensive digital representation and analysis.

- **Benefits:** Improved collaboration, accurate simulations, and better decision-making throughout the construction lifecycle.

## 14. Cybersecurity Measures:

- **Application:** Implementing robust cybersecurity measures to protect IoT devices and infrastructure from cyber threats.

- **Benefits:** Ensuring the integrity and security of data, preventing unauthorized access, and maintaining operational resilience.

## 15. Regulatory Compliance:

- **Application:** Adhering to relevant regulations and standards governing the use of IoT in construction and infrastructure.

- **Benefits:** Avoiding legal and compliance issues, ensuring the ethical use of data, and fostering public trust.

The integration of IoT in civil engineering and construction leads to the creation of intelligent, interconnected infrastructure that enhances efficiency, safety, and sustainability. However, it also requires careful consideration of data privacy, security, and the development of standardized practices to maximize the benefits of smart infrastructure.

## 8.2 Smart Cities

Smart Cities represent a paradigm shift in urban development, leveraging IoT (Internet of Things) and smart infrastructure to enhance efficiency, sustainability, and quality of life.

In the context of civil engineering and construction, the concept of smart cities involves the integration of advanced technologies to create intelligent, interconnected urban environments. Here are key aspects of smart cities with reference to IoT and smart infrastructure:

1. **IoT-enabled Infrastructure:**

   - **Description:** Smart cities leverage IoT to embed sensors, actuators, and communication systems into various infrastructure components such as roads, buildings, utilities, and transportation systems.

- **Benefits:** Real-time data collection and monitoring, predictive maintenance, and enhanced infrastructure management.

2. **Smart Buildings:**

- **Description:** Buildings in smart cities are equipped with IoT devices for energy management, security, and automation. Sensors monitor occupancy, lighting, and temperature, optimizing building operations.

- **Benefits:** Energy efficiency, improved occupant comfort, and streamlined building maintenance.

3. **Intelligent Transportation Systems (ITS):**

- **Description:** IoT is applied to create intelligent transportation networks, including smart traffic lights, connected vehicles, and real-time traffic monitoring.

- **Benefits:** Reduced congestion, enhanced traffic flow, and improved safety through data-driven traffic management.

4. **Smart Grids:**

- **Description:** IoT technologies are integrated into electrical grids to enable smart grid systems. This includes monitoring energy consumption, managing distributed energy resources, and optimizing power distribution.

- **Benefits:** Energy efficiency, better integration of renewable energy sources, and improved resilience of the power grid.

5. **Waste Management:**

- **Description:** IoT sensors are employed in waste bins to monitor fill levels. Smart waste management systems use data analytics to optimize collection routes and schedules.

- **Benefits:** Reduced operational costs, minimized environmental impact, and efficient waste collection.

6. **Water Management:**

- **Description:** IoT devices monitor water infrastructure, detect leaks, and optimize water distribution. Smart irrigation systems adjust based on weather conditions and soil moisture.

- **Benefits:** Water conservation, early leak detection, and improved water resource management.

7. **Public Safety and Security:**

- **Description:** Smart city infrastructure integrates IoT for public safety, including surveillance cameras, gunshot detection systems, and emergency response networks.

- **Benefits:** Enhanced situational awareness, rapid response to emergencies, and improved overall safety.

8. **Urban Mobility:**

- **Description:** IoT supports smart mobility solutions, including connected vehicles, intelligent parking systems, and real-time public transportation tracking.

- **Benefits:** Reduced traffic congestion, efficient parking management, and improved accessibility for residents.

9. **Environmental Monitoring:**

- **Description:** IoT sensors monitor air and water quality, noise levels, and other environmental parameters. Data is used to address pollution and enhance environmental sustainability.

- **Benefits:** Early detection of environmental issues, compliance with regulations, and informed decision-making for environmental policies.

10. **Citizen Engagement:**

- **Description:** Smart city initiatives leverage IoT for citizen engagement platforms, allowing residents to provide feedback, access services, and participate in decision-making.

- **Benefits:** Enhanced community involvement, improved service delivery, and a more responsive city government.

11. **Data Analytics and Predictive Insights:**

- **Description:** The data generated by IoT devices is analyzed to derive actionable insights. Predictive analytics helps in anticipating infrastructure issues and optimizing city operations.

- **Benefits:** Informed decision-making, proactive maintenance, and continuous improvement of urban services.

## 12. Interconnected Systems:

- **Description:** Integration of diverse IoT systems, creating a network where different components communicate and share data. For example, traffic data influencing building energy management.

- **Benefits:** Improved efficiency, holistic city planning, and a seamless experience for residents and businesses.

## 13. Resilience and Sustainability:

- **Description:** IoT-enabled infrastructure contributes to the resilience and sustainability of smart cities by optimizing resource use, reducing environmental impact, and improving disaster preparedness.

- **Benefits:** Enhanced city resilience, reduced ecological footprint, and long-term sustainability.

## 14. Challenges and Considerations:

- **Description:** Despite the benefits, smart cities face challenges related to data privacy, cybersecurity, standardization, and the need for robust governance models to ensure responsible and ethical use of technology.

- **Benefits:** Addressing challenges to build trust, ensure citizen privacy, and establish a framework for sustainable and inclusive smart city development.

In summary, the convergence of IoT and smart infrastructure in civil engineering and construction is a key driver in the evolution of smart cities. The integration of advanced technologies aims to create urban environments that are more sustainable, efficient, and responsive to the needs of their residents.

## 8.3 Sensors and Data Collection

Smart infrastructure relies on a variety of sensors and data collection methods to monitor, analyze, and optimize various components.

These sensors play a crucial role in gathering real-time information, enabling data-driven decision-making and enhancing the overall efficiency and performance of smart infrastructure. Here are some common types of sensors and their applications in smart infrastructure:

1. **Environmental Sensors:**

    - **Applications:** Monitoring air quality, detecting pollutants, measuring temperature, humidity, and noise levels.

- **Benefits:** Early detection of environmental issues, support for sustainable urban development, and improved public health.

2. **Structural Health Monitoring (SHM) Sensors:**

   - **Applications:** Monitoring the condition of structures such as bridges, buildings, and dams. Sensors measure factors like strain, vibration, and temperature.

   - **Benefits:** Early detection of structural issues, predictive maintenance, and improved safety.

3. **Smart Grid Sensors:**

   - **Applications:** Monitoring power distribution, measuring electricity consumption, detecting faults, and optimizing energy flow.

   - **Benefits:** Enhanced energy efficiency, reduced downtime, and improved reliability of power grids.

4. **Water Quality Sensors:**

   - **Applications:** Monitoring the quality of water in rivers, lakes, and reservoirs. Sensors measure parameters like pH, turbidity, and chemical concentrations.

   - **Benefits:** Early detection of water pollution, support for sustainable water management, and protection of water resources.

5. **Traffic and Transportation Sensors:**

   - **Applications:** Monitoring traffic flow, detecting vehicle presence, and optimizing traffic signal timings. Sensors include cameras, radar, and inductive loops.

   - **Benefits:** Reduced traffic congestion, improved road safety, and efficient transportation systems.

6. **Smart Parking Sensors:**

   - **Applications:** Monitoring parking space occupancy, guiding drivers to available parking spots, and optimizing parking management.

- **Benefits:** Reduced traffic congestion, improved parking efficiency, and enhanced urban mobility.

7. **Waste Management Sensors:**

   - **Applications:** Monitoring waste bin fill levels, optimizing waste collection routes, and managing recycling processes.

   - **Benefits:** Efficient waste collection, reduced operational costs, and improved sustainability.

8. **Gas and Leak Sensors:**

   - **Applications:** Detecting gas leaks in pipelines, monitoring industrial emissions, and ensuring workplace safety.

   - **Benefits:** Early detection of hazardous situations, prevention of environmental incidents, and improved safety protocols.

9. **Smart Lighting Sensors:**

   - **Applications:** Adjusting lighting levels based on occupancy and natural light conditions. Sensors include motion detectors and ambient light sensors.

   - **Benefits:** Energy efficiency, reduced electricity consumption, and improved lighting quality.

10. **Weather Sensors:**

    - **Applications:** Monitoring weather conditions, including temperature, humidity, wind speed, and precipitation.

    - **Benefits:** Improved weather forecasts, support for disaster preparedness, and optimization of infrastructure resilience.

11. **IoT-enabled Cameras:**

    - **Applications:** Surveillance, traffic monitoring, and public safety. Cameras may incorporate video analytics for advanced functionalities.

    - **Benefits:** Enhanced security, real-time monitoring, and data-driven insights for law enforcement and urban planning.

12. **GPS and Location Sensors:**

- **Applications:** Tracking the movement of vehicles, pedestrians, and assets. Used in navigation systems and location-based services.

- **Benefits:** Improved transportation efficiency, accurate mapping, and support for location-aware applications.

## 13. Vibration Sensors:

- **Applications:** Monitoring vibrations in structures, machinery, and equipment. Used for predictive maintenance and structural health assessments.

- **Benefits:** Early detection of mechanical issues, prevention of equipment failures, and improved reliability.

## 14. Asset Tracking Sensors:

- **Applications:** Tracking the location and status of infrastructure assets such as equipment, vehicles, and tools.

- **Benefits:** Improved asset management, reduced loss or theft, and optimization of resource utilization.

Data collected from these sensors contribute to a broader IoT ecosystem, where interconnected devices share information to create a comprehensive view of smart infrastructure. This data is then analyzed to derive actionable insights, optimize operations, and enhance the overall performance of urban environments.

## 8.4 Communication Networks

Communication networks are a critical component of IoT (Internet of Things) and smart infrastructure in civil engineering and construction.

These networks facilitate the seamless exchange of data between various sensors, devices, and systems, enabling real-time monitoring, control, and optimization of infrastructure components. Here are key aspects related to communication networks in the context of IoT and smart infrastructure:

1. **Wireless Sensor Networks (WSN):**

   - **Description:** WSNs consist of spatially distributed sensors that communicate wirelessly. They play a vital role in collecting data from sensors embedded in infrastructure components.

- **Applications:** Structural health monitoring, environmental sensing, and asset tracking.

- **Benefits:** Reduced wiring complexity, flexibility in sensor placement, and cost-effective data collection.

2. **5G Technology:**

   - **Description:** 5G is the fifth generation of mobile networks, offering high data rates, low latency, and increased connectivity. It provides a robust foundation for IoT applications.

   - **Applications:** High-bandwidth applications, real-time communication, and support for a massive number of connected devices.

   - **Benefits:** Faster data transfer, low latency for critical applications, and improved network capacity.

3. **Low-Power Wide-Area Networks (LPWAN):**

   - **Description:** LPWAN technologies, such as LoRaWAN and NB-IoT, are designed for low-power, long-range communication. They are suitable for IoT devices with low data-rate requirements.

   - **Applications:** Remote monitoring, smart agriculture, and low-power IoT sensors.

   - **Benefits:** Extended battery life for devices, long-range connectivity, and cost-effective deployment.

4. **Mesh Networks:**

   - **Description:** Mesh networks consist of interconnected devices that can relay data to extend the network's reach. They offer redundancy and flexibility in communication paths.

   - **Applications:** Smart lighting systems, smart cities, and large-scale sensor deployments.

   - **Benefits:** Improved reliability, self-healing capabilities, and scalability for diverse deployments.

5. **Satellite Communication:**

- **Description:** Satellite communication provides connectivity in remote or challenging environments where traditional networks may be unavailable.

- **Applications:** Remote infrastructure monitoring, disaster response, and global asset tracking.

- **Benefits:** Wide coverage, global reach, and resilience in areas with limited terrestrial network infrastructure.

6. **Fiber Optic Networks:**

- **Description:** Fiber optic networks offer high-speed, reliable communication through the transmission of data via optical fibers. They are often used as backbones in smart city infrastructures.

- **Applications:** High-bandwidth applications, such as video surveillance and data-intensive operations.

- **Benefits:** High data transfer rates, low latency, and resistance to electromagnetic interference.

7. **Edge Computing:**

- **Description:** Edge computing involves processing data near the source of generation rather than relying solely on centralized cloud servers. It enhances real-time processing for IoT applications.

- **Applications:** Local data analytics, reduced latency for critical applications, and bandwidth optimization.

- **Benefits:** Lower latency, reduced data transfer requirements, and improved responsiveness.

8. **Cybersecurity Protocols:**

- **Description:** Cybersecurity protocols and measures are essential to protect data transmitted across IoT devices and networks.

- **Applications:** Secure communication, data encryption, and protection against cyber threats.

- **Benefits:** Safeguarding sensitive data, maintaining the integrity of communications, and ensuring privacy.

9. **Integration with Smart Grids:**

- **Description:** Integration with smart grids involves communication between IoT devices in the energy sector. This facilitates efficient energy distribution and consumption.

- **Applications:** Demand response, grid monitoring, and integration of renewable energy sources.

- **Benefits:** Optimized energy usage, reduced energy wastage, and enhanced grid resilience.

10. **Interoperability Standards:**

- **Description:** Standardization of communication protocols ensures interoperability among diverse IoT devices and systems, promoting seamless integration.

- **Applications:** Cross-vendor compatibility, ease of integration, and future-proofing.

- **Benefits:** Facilitates a unified ecosystem, reduces deployment complexities, and supports scalability.

Effective communication networks form the backbone of smart infrastructure, enabling the integration and coordination of diverse IoT devices and systems. The choice of communication technologies depends on specific project requirements, scalability considerations, and the nature of the applications implemented in civil engineering and construction projects.

## 8.5 AI-driven Infrastructure Management

AI-driven infrastructure management represents a transformative approach to efficiently design, build, and maintain infrastructure.

Here are key aspects of AI-driven infrastructure management:

1. **Data-Driven Decision-Making:**

   - **Description:** AI leverages data analytics to process vast amounts of information from various sources, including sensors, historical data, and real-time monitoring systems.

   - **Applications:** Informed decision-making in project planning, resource allocation, risk assessment, and maintenance strategies.

- **Benefits:** Improved accuracy, reduced reliance on assumptions, and enhanced decision-making capabilities.

2. **Predictive Analytics:**

   - **Description:** Predictive analytics models use AI algorithms to forecast future outcomes based on historical and real-time data. This enables proactive planning and risk mitigation.

   - **Applications:** Forecasting project timelines, identifying potential issues, and optimizing resource allocation.

   - **Benefits:** Anticipating challenges, preventing delays, and improving overall project efficiency.

3. **Asset Performance Management:**

   - **Description:** AI monitors and assesses the performance of infrastructure assets throughout their lifecycle. It includes predictive maintenance, condition monitoring, and risk analysis.

   - **Applications:** Predicting equipment failures, optimizing maintenance schedules, and extending the lifespan of assets.

   - **Benefits:** Minimized downtime, reduced maintenance costs, and improved overall asset performance.

4. **Smart Infrastructure Monitoring:**

   - **Description:** Smart sensors and IoT devices continuously monitor the condition of infrastructure components. AI processes the data to detect anomalies, structural issues, or potential risks.

   - **Applications:** Structural health monitoring, real-time anomaly detection, and early warning systems.

   - **Benefits:** Enhanced safety, timely intervention, and improved infrastructure resilience.

5. **Automated Project Planning and Scheduling:**

   - **Description:** AI optimizes project planning and scheduling by analyzing historical data, identifying patterns, and generating efficient schedules.

- **Applications:** Efficient resource allocation, task sequencing, and timeline optimization.

- **Benefits:** Reduced project delays, improved resource utilization, and streamlined workflows.

6. **Supply Chain Optimization:**

   - **Description:** AI analyzes data related to the supply chain, predicting material requirements, managing inventory, and optimizing logistics.

   - **Applications:** Efficient procurement, reduced costs, and timely delivery of construction materials.

   - **Benefits:** Cost savings, minimized delays, and improved overall project efficiency.

7. **Human-Machine Collaboration (Augmented Intelligence):**

   - **Description:** AI collaborates with human experts, augmenting their capabilities. It involves leveraging AI for data analysis, leaving more time for strategic decision-making.

   - **Applications:** Collaborative problem-solving, data interpretation, and strategic decision support.

   - **Benefits:** Enhanced productivity, improved decision quality, and a synergy between human intuition and AI analytics.

8. **Energy Efficiency and Sustainability:**

   - **Description:** AI contributes to sustainability by analyzing data related to energy consumption, waste management, and environmental impact.

   - **Applications:** Designing eco-friendly structures, implementing energy-efficient solutions, and optimizing resource usage.

   - **Benefits:** Meeting sustainability goals, reducing environmental impact, and aligning with green construction practices.

9. **Continuous Learning and Adaptability:**

   - **Description:** AI systems continuously learn and adapt based on new data and experiences. This fosters a culture of continuous improvement in infrastructure management.

- **Applications:** Learning from project outcomes, adapting to changing conditions, and staying updated on industry trends.

- **Benefits:** Improved system performance, adaptability to evolving challenges, and staying at the forefront of technological advancements.

## 10. Ethical Considerations and Governance:

- **Description:** Addressing ethical considerations involves establishing governance frameworks to guide the responsible deployment of AI in infrastructure management.

- **Applications:** Mitigating biases, ensuring data privacy, and promoting transparency in decision-making processes.

- **Benefits:** Building trust, preventing unintended consequences, and ensuring responsible AI integration.

AI-driven infrastructure management in civil engineering and construction holds the potential to revolutionize the industry by optimizing processes, improving decision-making, and enhancing the overall sustainability and resilience of infrastructure projects.

# 8.6 Real-world Examples / Use Cases

1. **Smart Infrastructure and IoT in Civil Engineering and Construction:**

   A) **Smart Concrete Monitoring (Company: Giatec):**

      - Giatec provides smart concrete monitoring solutions that utilize IoT devices embedded in concrete structures. These devices monitor the curing and strength development of concrete in real-time, optimizing construction processes.

   B) **Smart Buildings (Company: Siemens):**

      - Siemens offers smart building solutions that integrate IoT in civil engineering and construction. Their systems use sensors and connected devices to enhance building efficiency, comfort, and energy management.

2. **Internet of Things (IoT) in Civil Engineering and Construction:**

   C) **Construction Site Monitoring (Company: Pillar Technologies - a division of Gilbane):**

      - Pillar Technologies, a division of Gilbane, utilizes IoT for construction site monitoring. The platform deploys sensors to collect data on environmental conditions, safety, and risk factors, enhancing overall project management.

   D) **Equipment Tracking (Company: Caterpillar - Cat Connect):**

      - Caterpillar's Cat Connect utilizes IoT for equipment tracking in construction. IoT-enabled sensors on machinery collect data on usage, performance, and maintenance needs, optimizing equipment management.

3. **Smart Cities Smart Infrastructure and IoT in Civil Engineering and Construction:**

   E) **Smart Street Lighting (City: Barcelona):**

      - Barcelona implemented smart street lighting as part of its smart city initiatives. IoT-connected streetlights adjust brightness based on real-time data, optimizing energy usage and enhancing city infrastructure.

### F) Integrated Traffic Management (City: Singapore):

- Singapore has embraced IoT for integrated traffic management. Smart sensors and cameras collect real-time traffic data, enabling dynamic traffic control to improve congestion and enhance transportation infrastructure.

## 4. Sensors and Data Collection Smart Infrastructure and IoT in Civil Engineering and Construction:

### G) Wireless Sensor Networks (Company: Libelium):

- Libelium provides wireless sensor networks for civil engineering and construction. These sensors collect data on environmental conditions, structural health, and other factors, supporting real-time monitoring and decision-making.

### H) Remote Monitoring of Structural Health (Company: Sisgeo):

- Sisgeo offers solutions for remote monitoring of structural health using IoT. Their sensors collect data on deformation, settlement, and vibrations, enabling continuous monitoring of infrastructure integrity.

## 5. Communication Networks for Smart Infrastructure and IoT in Civil Engineering and Construction:

### I) LPWAN for Smart Cities (Company: Semtech - LoRa):

- Semtech's LoRa (Low Power Wide Area Network) technology provides communication networks for smart cities. It facilitates long-range, low-power communication for IoT devices, supporting various applications in civil engineering and construction.

### J) 5G-enabled Construction Sites (Company: Ericsson):

- Ericsson contributes to 5G-enabled construction sites, providing high-speed, low-latency communication. This enables real-time data exchange, supporting IoT applications for improved construction site efficiency.

6. **AI-driven Infrastructure Management in Civil Engineering and Construction:**

   **K) AI-driven Asset Management (Company: PlanGrid - Autodesk):**

   - PlanGrid, a part of Autodesk, incorporates AI-driven asset management in construction. The platform uses AI algorithms to analyze project data, enhancing asset tracking and management for construction projects.

   **L) Predictive Maintenance in Infrastructure (Company: Worldsensing):**

   - Worldsensing applies AI for predictive maintenance in infrastructure. Their platform analyzes IoT-collected data to predict and optimize maintenance schedules, ensuring the reliability of critical infrastructure components.

## 8.7 Chapter Summary: Key Points

1. Smart infrastructure and IoT integration involve embedding sensors and communication technologies for real-time monitoring and control in civil engineering, fostering efficiency and sustainability.

2. Structural health monitoring through IoT allows early issue detection and preventive maintenance, ensuring heightened safety and longevity of infrastructure.

3. Smart transportation systems optimize traffic flow, reduce congestion, and enhance overall transportation planning through IoT technologies.

4. IoT-driven construction sites utilize real-time data for improved safety, resource allocation, and project timeline efficiency.

5. Environmental monitoring with IoT measures air and water quality, supporting compliance and sustainable construction practices.

6. IoT enhances asset management by tracking and monitoring infrastructure components, predicting maintenance needs and optimizing lifecycle management.

7. Energy efficiency in smart infrastructure is achieved through IoT-enabled monitoring and control of energy consumption.

8. IoT aids emergency response and disaster management by providing early warnings and minimizing infrastructure and safety impact.

9. IoT platforms and data analytics derive actionable insights, informing decision-making for efficient resource allocation and proactive maintenance.

10. Key functions of smart infrastructure and IoT include structural health monitoring, asset tracking, energy management, and environmental monitoring.

11. Smart cities leverage IoT to create interconnected urban environments, enhancing efficiency, sustainability, and quality of life.

12. IoT-enabled smart buildings optimize energy usage, enhance occupant comfort, and streamline building operations.

13. Intelligent transportation systems utilize IoT for reduced congestion, improved traffic flow, and enhanced safety.

14. IoT contributes to waste and water management efficiency in smart cities, optimizing collection routes and distribution.

15. Challenges in smart cities include data privacy, cybersecurity, and the need for governance models to ensure responsible technology use.

## 8.8 Concept Check: Q&A Sessions

1.  What is the key role of IoT in civil engineering and construction practices, particularly in the context of smart infrastructure?

2.  How does structural health monitoring utilize IoT in civil engineering, and what benefits does it offer?

3.  What are the key functions of IoT in smart construction sites, and how do they contribute to project efficiency and safety?

4.  How does IoT contribute to energy efficiency in smart infrastructure, and what areas does it monitor and control?

5.  What is the significance of IoT in emergency response and disaster management within the field of civil engineering?

6.  How does the integration of IoT in asset management benefit civil engineering, and what types of infrastructure components does it track and monitor?

7.  What are the key applications of IoT in traffic management, and how does it improve transportation infrastructure efficiency?

8.  How does IoT contribute to environmental monitoring in civil engineering, and what types of environmental factors can be measured?

9.  What is the role of data analytics in smart infrastructure, and how is data collected from IoT devices processed to derive actionable insights?

10. How does AI-driven infrastructure management utilize predictive analytics, and what benefits does it offer in terms of project planning and risk mitigation?

# 9.0 AI in Quality Control and Assurance

Artificial Intelligence (AI) is revolutionizing quality control and assurance processes in civil engineering and construction by providing innovative solutions that enhance efficiency, accuracy, and reliability.

One significant application of AI in this context is the use of computer vision for automated inspection. AI-powered systems can analyze images and videos captured by cameras and drones to detect defects, anomalies, and deviations in construction materials and structures.

Artificial Intelligence (AI) serves as a transformative force in quality control and assurance within civil engineering and construction, introducing nine key functions to enhance precision, efficiency, and proactive issue identification. These functions include the utilization of

AI-powered computer vision systems and drones for automated inspection, enabling enhanced speed and accuracy in defect identification.

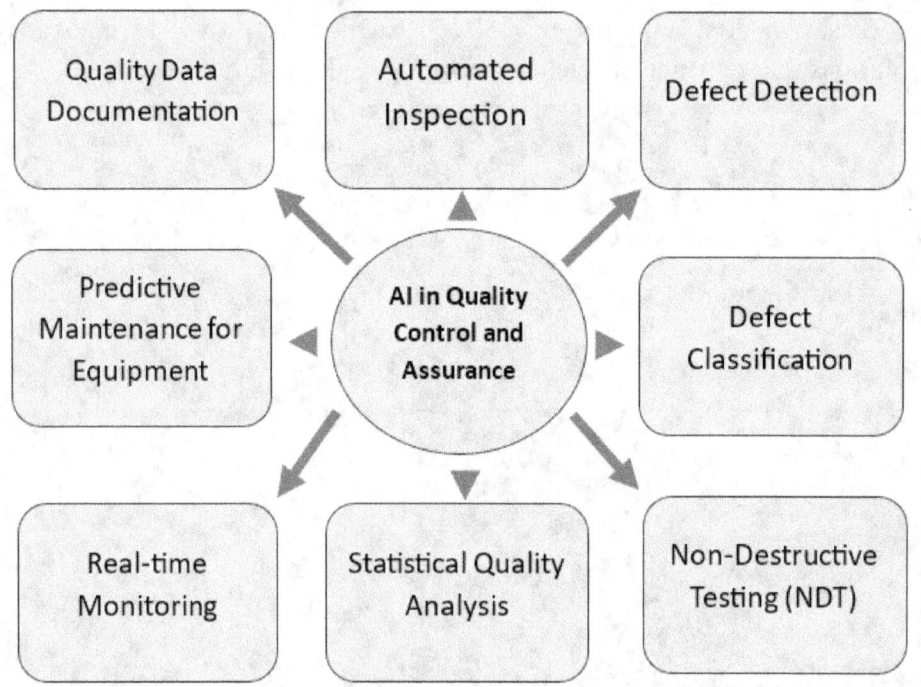

AI algorithms analyze images and sensor data to detect and classify defects, ensuring early identification and adherence to quality standards. AI-based systems assess construction materials' quality, promoting consistency and reducing the risk of substandard construction. Integration of AI in non-destructive testing techniques allows for precise evaluation of structural integrity without causing damage. Statistical quality analysis through AI-driven data analytics identifies patterns and trends in construction processes, enabling continuous improvement. Real-time monitoring, facilitated by AI-based sensors, ensures immediate detection of deviations from quality standards. Predictive maintenance for equipment, enabled by AI, minimizes downtime and extends equipment lifespan. AI also automates the documentation of quality-related data, simplifying auditing processes and enhancing transparency. Together, these functions empower the construction industry to deliver higher-quality projects, reduce rework, and uphold stringent quality standards, ultimately ensuring successful and durable project completion.

Machine learning algorithms play a pivotal role in predicting and preventing construction defects. By analyzing historical data and learning from past projects, AI models can identify patterns and potential issues early in the construction process. This proactive approach allows for the implementation of preventive measures and quality improvements.

AI-driven robotics contribute to quality control by performing tasks that are repetitive, labor-intensive, or hazardous for humans. Robotic systems equipped with sensors and AI algorithms can conduct inspections, measurements, and material testing with precision, reducing the risk of errors and enhancing the overall quality of construction projects.

Predictive analytics is another aspect where AI excels in quality control. By analyzing data from various sources, including project plans, weather conditions, and historical performance, AI models can predict potential risks and variations in construction quality. This foresight allows project managers to make informed decisions and allocate resources effectively.

Smart sensors embedded in construction materials and structures enable real-time monitoring of key parameters such as temperature, moisture, and stress. AI algorithms analyze the data generated by these sensors to assess the quality of materials and identify any deviations from specifications, facilitating timely interventions and quality assurance.

Natural Language Processing (NLP) is applied in AI systems to process and analyze textual data, such as construction documents and reports. NLP algorithms can extract valuable insights, identify key quality indicators, and ensure that construction projects adhere to relevant standards and regulations.

The integration of AI in quality control contributes to the development of digital twins in construction. Digital twins are virtual replicas of physical structures that allow for real-time monitoring and analysis. AI algorithms work in conjunction with digital twins to simulate various scenarios, assess potential risks, and optimize construction processes for quality assurance.

AI technologies facilitate non-destructive testing methods in civil engineering. Ultrasonic testing, thermal imaging, and acoustic monitoring, powered by AI, enable engineers to assess the integrity of structures without causing damage. This enhances the overall quality control process and ensures the safety and durability of construction projects.

Quality data management is streamlined with AI solutions, enabling efficient storage, retrieval, and analysis of vast amounts of construction data. This contributes to a centralized and organized approach to quality control, making it easier for stakeholders to access relevant information and make data-driven decisions.

The adoption of AI in quality control and assurance reflects a transformative shift in the construction industry, promoting proactive measures, precision, and adherence to quality standards. As AI technologies continue to advance, their integration into quality control processes will likely become even more sophisticated, further elevating the standards of excellence in civil engineering and construction.

## 9.1 Automated Inspection

Automated inspection plays a crucial role in quality control and assurance within civil engineering and construction, leveraging AI technologies to enhance efficiency and accuracy.

Here are key aspects of AI-driven automated inspection in this context:

1. **Computer Vision and Image Analysis:**

   - **Description:** AI algorithms, particularly computer vision, analyze visual data from images and videos captured on construction sites.

- **Applications:** Detection of defects, evaluation of construction components, and assessment of adherence to design specifications.

- **Benefits:** Rapid and accurate identification of issues, reducing reliance on manual inspections and improving overall quality control.

2. **Defect Detection and Classification:**

    - **Description:** AI models are trained to identify and classify defects or deviations from project specifications.

    - **Applications:** Detecting cracks, structural deformities, or deviations in materials during construction processes.

    - **Benefits:** Early identification of defects, prevention of costly rework, and improved adherence to quality standards.

3. **Automated Drone Inspections:**

    - **Description:** Drones equipped with AI-powered cameras perform aerial inspections of construction sites.

    - **Applications:** Monitoring large areas, capturing high-resolution images, and identifying issues in hard-to-reach or hazardous locations.

    - **Benefits:** Enhanced safety, efficient data collection, and comprehensive site coverage.

4. **3D Laser Scanning:**

    - **Description:** AI processes data from 3D laser scans of construction projects, creating detailed models for analysis.

    - **Applications:** Precise measurements, identification of discrepancies, and comparison with design specifications.

    - **Benefits:** Accurate documentation, improved clash detection, and streamlined quality assurance processes.

5. **Automated Data Collection from Sensors:**

    - **Description:** Sensors embedded in construction materials or structures collect real-time data, which is analyzed by AI.

- **Applications:** Monitoring structural health, tracking environmental conditions, and identifying potential issues.

- **Benefits:** Early detection of anomalies, proactive maintenance, and improved overall quality assurance.

6. **Machine Learning for Pattern Recognition:**

    - **Description:** Machine learning algorithms analyze data patterns to identify irregularities or deviations from expected norms.

    - **Applications:** Recognizing patterns in construction processes, materials, or structures to flag potential quality issues.

    - **Benefits:** Improved accuracy in defect identification, reduced false positives, and efficient processing of large datasets.

7. **Integration with Building Information Modeling (BIM):**

    - **Description:** AI is integrated with BIM systems for a comprehensive approach to quality control and assurance.

    - **Applications:** Cross-referencing as-built structures with BIM models, identifying discrepancies, and ensuring design compliance.

    - **Benefits:** Enhanced collaboration, improved accuracy in construction documentation, and streamlined quality control workflows.

8. **Real-time Monitoring and Alerts:**

    - **Description:** AI systems monitor construction processes in real-time, generating alerts for deviations from quality standards.

    - **Applications:** Immediate notification of issues, enabling prompt corrective actions and preventing potential defects.

    - **Benefits:** Timely interventions, reduced rework, and improved overall construction quality.

9. **Data Integration from Multiple Sources:**

    - **Description:** AI consolidates data from various sources, including sensors, drones, and manual inspections.

- **Applications:** Comprehensive analysis, cross-referencing data points, and providing a holistic view of construction quality.

- **Benefits:** Improved decision-making, better overall project understanding, and enhanced quality assurance.

## 10. Continuous Improvement through Feedback Loop:

- **Description:** AI systems utilize feedback from inspections to continuously improve their defect recognition capabilities.

- **Applications:** Learning from past projects, refining algorithms, and adapting to evolving construction requirements.

- **Benefits:** Enhanced accuracy over time, increased efficiency in defect detection, and improved overall quality control processes.

Automated inspection powered by AI in quality control and assurance offers a proactive and efficient approach to ensuring construction projects meet or exceed quality standards. This technology not only improves accuracy but also contributes to the overall success and longevity of infrastructure projects.

## 9.2 Quality Assurance Algorithms

Quality assurance algorithms in the context of AI-driven quality control and assurance in civil engineering and construction involve sophisticated methodologies to ensure the adherence to quality standards, identify defects, and optimize construction processes.

Here are key aspects of quality assurance algorithms in this domain:

1.  **Defect Detection Algorithms:**

    -   **Description:** Algorithms designed to identify defects or deviations from design specifications.

    -   **Applications:** Identifying cracks, deformities, or inconsistencies in construction materials or structures.

- **Benefits:** Early detection of defects, minimizing rework, and ensuring compliance with quality standards.

2. **Computer Vision for Visual Inspection:**

   - **Description:** Computer vision algorithms analyze visual data from images or videos to assess the visual quality of construction components.

   - **Applications:** Visual inspection of surfaces, finishes, and overall construction quality.

   - **Benefits:** Objective assessment, rapid inspection, and improved accuracy in identifying visual defects.

3. **Pattern Recognition in Construction Processes:**

   - **Description:** Machine learning algorithms recognize patterns in construction processes to identify irregularities.

   - **Applications:** Detecting deviations in construction workflows, materials application, or assembly processes.

   - **Benefits:** Improved process efficiency, early identification of anomalies, and optimization of construction procedures.

4. **Materials Compliance Algorithms:**

   - **Description:** Algorithms that assess the compliance of construction materials with specified standards.

   - **Applications:** Verifying the quality and composition of materials used in construction.

   - **Benefits:** Ensuring materials meet required standards, preventing the use of substandard components.

5. **Structural Integrity Assessment Algorithms:**

   - **Description:** Algorithms designed to assess the structural integrity of constructed elements.

   - **Applications:** Evaluating the stability and safety of structures during and after construction.

   - **Benefits:** Ensuring structural safety, identifying potential issues, and optimizing construction practices.

6. **Real-time Monitoring and Alerts:**

- **Description:** Algorithms that enable real-time monitoring of construction processes and generate alerts for deviations.

- **Applications:** Immediate notification of quality issues, allowing prompt corrective actions.

- **Benefits:** Timely interventions, reducing the likelihood of defects, and ensuring continuous quality improvement.

7. **Statistical Quality Control Algorithms:**

- **Description:** Statistical algorithms analyze data to determine if construction processes are within acceptable statistical limits.

- **Applications:** Statistical analysis of construction data to identify variations and trends.

- **Benefits:** Ensuring consistency and reliability in construction processes, reducing variability in quality.

8. **Machine Learning for Dynamic Quality Models:**

- **Description:** Machine learning models that dynamically adapt to changing construction conditions and requirements.

- **Applications:** Adapting quality control models based on evolving project specifications.

- **Benefits:** Improved accuracy over time, flexibility in responding to changing conditions, and continuous improvement.

9. **Integration with Building Information Modeling (BIM):**

- **Description:** Algorithms integrated with BIM systems for cross-referencing constructed elements with design specifications.

- **Applications:** Verifying that the as-built structures align with BIM models.

- **Benefits:** Improved collaboration, enhanced accuracy in construction documentation, and streamlined quality control workflows.

### 10. Automated Documentation and Reporting:

- **Description:** Algorithms that automate the generation of reports summarizing quality control findings.

- **Applications:** Creating comprehensive reports on construction quality for stakeholders.

- **Benefits:** Efficient documentation, transparency in quality assessment, and improved communication among project participants.

These quality assurance algorithms collectively contribute to an advanced and proactive approach to quality control in civil engineering and construction. By leveraging AI technologies, construction projects can achieve higher levels of accuracy, consistency, and compliance with quality standards throughout various stages of development.

## 9.3 Non-Destructive Testing

Non-Destructive Testing (NDT) techniques play a crucial role in quality control and assurance in civil engineering and construction, particularly when integrated with artificial intelligence (AI).

Non-Destructive Testing (NDT) is a crucial set of techniques employed to evaluate the properties of materials and structures without causing any harm or alteration to the tested components. The primary objective of NDT is to identify defects, anomalies, or structural weaknesses in a non-intrusive manner, ensuring the integrity and safety of various materials used in industries such as manufacturing, construction, and aerospace. By utilizing NDT methods, professionals can assess the quality and reliability of materials, welds, and components, ultimately preventing potential failures and ensuring compliance with safety standards.

Traditional NDT methods encompass a diverse range of techniques that have proven effective in different scenarios. Ultrasonic testing involves the use of high-frequency sound waves to detect internal flaws, while radiographic testing employs X-rays or gamma rays to inspect the internal structure of materials. Magnetic particle testing identifies surface and near-surface defects by applying a magnetic field and observing particle accumulations. Visual inspection, a fundamental method, relies on direct observation to identify surface irregularities. Each of these methods serves a specific purpose, collectively contributing to a comprehensive evaluation of materials and structures, and they are widely employed across industries to maintain the structural integrity and safety of critical components.

Here are key aspects of how AI can enhance NDT for quality control and assurance:

a. **Integration of AI in NDT:**

- **Description:** AI is employed to enhance the analysis, interpretation, and decision-making processes involved in NDT.

- **Applications:** Improving the accuracy of defect detection, automating analysis, and providing real-time insights.

b. **Automated Defect Detection:**

- **Description:** AI algorithms are trained to automatically identify and classify defects in materials or structures.

- **Applications:** Detecting cracks, voids, corrosion, or other anomalies during NDT inspections.

- **Benefits:** Faster and more accurate defect identification, reducing the risk of human error.

c. **Machine Learning for Pattern Recognition:**

- **Description:** Machine learning models learn from NDT data patterns to recognize subtle defects.

- **Applications:** Recognizing complex patterns indicative of structural issues or material defects.

- **Benefits:** Improved sensitivity to nuanced defects, enhanced predictive capabilities.

d. **Real-time Analysis and Decision Support:**

- **Description:** AI enables real-time analysis of NDT data, providing instant insights.

- **Applications:** On-the-fly decision-making during inspections, especially in critical or time-sensitive situations.

- **Benefits:** Immediate identification of issues, facilitating timely corrective actions.

e. **Data Fusion for Comprehensive Assessment:**

- **Description:** AI integrates data from multiple NDT methods to provide a comprehensive assessment.

- **Applications:** Combining data from ultrasonic, radiographic, and other NDT techniques for a holistic view.

- **Benefits:** Enhanced accuracy and reliability by considering multiple sources of information.

f. **Predictive Maintenance Modeling:**

- **Description:** AI models predict the future condition of structures based on historical NDT data.

- **Applications:** Forecasting maintenance needs and prioritizing interventions based on structural health.

- **Benefits:** Proactive maintenance, extending the lifespan of structures and minimizing downtime.

g. **Quantitative Analysis with AI Algorithms:**

- **Description:** AI algorithms provide quantitative analysis of NDT data, going beyond qualitative assessments.

- **Applications:** Measuring the extent and severity of defects, enabling precise evaluations.

- **Benefits:** Improved accuracy in assessing defect dimensions and characteristics.

h. **Integration with Structural Health Monitoring (SHM):**

- **Description:** AI-enhanced NDT is integrated into SHM systems for continuous monitoring.

- **Applications:** Combining periodic NDT inspections with ongoing structural health monitoring.

- **Benefits:** Comprehensive and continuous monitoring of structural integrity.

i. **Edge Computing for In-situ Analysis:**

- **Description:** AI algorithms are deployed on-site using edge computing for immediate analysis.

- **Applications:** Conducting NDT analysis directly at the inspection site.

- **Benefits:** Reduced latency, enabling rapid decision-making without relying on centralized processing.

j. **Automated Reporting and Documentation:**

- **Description:** AI automates the generation of reports summarizing NDT findings.

- **Applications:** Efficiently documenting and communicating NDT results to stakeholders.

- **Benefits:** Streamlined reporting processes, facilitating collaboration and decision-making.

The integration of AI with NDT in civil engineering and construction significantly enhances the efficiency, accuracy, and reliability of quality control and assurance processes. This integration enables a more proactive and data-driven approach to ensuring the integrity and safety of structures throughout their lifecycle.

## 9.4 Certification and Compliance

Certification and compliance play a vital role in ensuring the responsible and ethical use of AI in quality control and assurance in civil engineering and construction.

Here are key aspects related to certification and compliance in this context:

1. **Industry Standards and Regulations:**

   - **Description:** Certification processes often align with established industry standards and regulations.

- **Applications:** Adhering to standards such as ISO 9001 (Quality Management), ISO 19650 (Building Information Modeling), and other relevant guidelines.

- **Benefits:** Ensures that AI applications in quality control and assurance meet recognized benchmarks for safety and effectiveness.

2. **Certification Bodies and Authorities:**

   - **Description:** Independent bodies and regulatory authorities may provide certifications for AI applications.

   - **Applications:** Seeking certifications from recognized organizations that specialize in AI ethics and safety.

   - **Benefits:** Enhances credibility and demonstrates a commitment to ethical AI practices.

3. **Ethical AI Frameworks:**

   - **Description:** Certification may involve adherence to ethical frameworks for AI development and deployment.

   - **Applications:** Incorporating principles from frameworks like the IEEE Ethically Aligned Design or AI ethics guidelines from reputable organizations.

   - **Benefits:** Addresses ethical considerations and guides the responsible use of AI in quality control.

4. **Algorithmic Transparency and Explainability:**

   - **Description:** Certification processes may require transparency and explainability in AI algorithms.

   - **Applications:** Implementing algorithms that can be understood and interpreted, especially in critical applications.

   - **Benefits:** Enables stakeholders to comprehend AI-driven decisions and fosters trust in the technology.

5. **Data Privacy and Security Compliance:**

   - **Description:** Certification should address compliance with data privacy and security regulations.

- **Applications:** Adhering to regulations like GDPR (General Data Protection Regulation) and implementing robust security measures.

- **Benefits:** Protects sensitive project data and ensures the privacy of individuals involved in construction projects.

6. **Bias Mitigation and Fairness:**

   - **Description:** Certification processes may require measures to mitigate biases in AI algorithms.

   - **Applications:** Implementing techniques to identify and address biases, ensuring fairness in decision-making.

   - **Benefits:** Minimizes the risk of discriminatory outcomes and promotes fairness in quality control processes.

7. **Human-AI Collaboration Standards:**

   - **Description:** Certification may involve standards for collaboration between humans and AI systems.

   - **Applications:** Defining protocols for human oversight, intervention, and collaboration in quality control activities.

   - **Benefits:** Ensures that AI complements human expertise rather than replacing it entirely.

8. **Continuous Monitoring and Auditing:**

   - **Description:** Certification processes may require ongoing monitoring and auditing of AI systems.

   - **Applications:** Implementing mechanisms for continuous monitoring, regular audits, and updates to AI models.

   - **Benefits:** Ensures that AI systems remain in compliance and effective over time.

9. **Legal and Liability Considerations:**

   - **Description:** Certification should address legal aspects and liabilities associated with AI applications.

   - **Applications:** Understanding and complying with legal frameworks related to AI in construction.

- **Benefits:** Clarifies responsibilities and liabilities, reducing legal risks associated with AI implementation.

10. **User Training and Awareness:**

- **Description:** Certification may involve training users and stakeholders on AI systems.

- **Applications:** Providing education and training programs to construction professionals interacting with AI in quality control.

- **Benefits:** Enhances awareness, usability, and acceptance of AI technologies in construction.

Certification and compliance frameworks are essential components of the responsible integration of AI in quality control and assurance in civil engineering and construction. By aligning with industry standards, ethical principles, and regulatory requirements, construction stakeholders can ensure the effective, safe, and ethical use of AI technologies.

# 9.5 Continuous Improvement

Continuous improvement is crucial in the application of AI in quality control and assurance in civil engineering and construction.

Here are key considerations for achieving continuous improvement in this context:

1. **Feedback Loops and Learning Systems:**

   - **Description:** Establish feedback loops within AI systems to continuously learn from data and outcomes.

   - **Applications:** Implement machine learning models that adapt and improve based on ongoing feedback from construction projects.

- **Benefits:** Enhances the accuracy and effectiveness of AI algorithms over time.

2. **Performance Monitoring and Metrics:**

   - **Description:** Define key performance indicators (KPIs) and metrics to assess the performance of AI in quality control.

   - **Applications:** Regularly monitor and analyze AI-driven processes, comparing outcomes against established benchmarks.

   - **Benefits:** Provides insights into the effectiveness of AI applications and identifies areas for improvement.

3. **Collaboration with Construction Professionals:**

   - **Description:** Foster collaboration between AI developers and construction professionals for real-world insights.

   - **Applications:** Engage construction teams to provide input, validate AI results, and suggest improvements based on practical experience.

   - **Benefits:** Ensures that AI solutions align with industry needs and challenges.

4. **Adaptive Algorithms and Model Iteration:**

   - **Description:** Develop algorithms that can adapt and iterate based on evolving project requirements.

   - **Applications:** Regularly update AI models to incorporate new data, construction methodologies, and quality standards.

   - **Benefits:** Keeps AI models relevant and effective in dynamic construction environments.

5. **Benchmarking Against Industry Standards:**

   - **Description:** Benchmark AI applications against industry standards and best practices.

   - **Applications:** Regularly assess AI systems against established benchmarks and standards in quality control and assurance.

- **Benefits:** Identifies areas where AI performance can be enhanced to meet or exceed industry expectations.

6. **User Training and Skill Development:**

   - **Description:** Invest in training programs to enhance the skills of construction professionals interacting with AI.

   - **Applications:** Provide ongoing training to ensure that users understand and leverage AI capabilities effectively.

   - **Benefits:** Improves user proficiency, leading to better collaboration and utilization of AI tools.

7. **Integration with BIM and Project Management Systems:**

   - **Description:** Ensure seamless integration with Building Information Modeling (BIM) and project management systems.

   - **Applications:** Facilitate data exchange and interoperability between AI systems and other construction management tools.

   - **Benefits:** Enhances the overall efficiency of project workflows and data utilization.

8. **Regular System Audits and Assessments:**

   - **Description:** Conduct regular audits and assessments of AI systems.

   - **Applications:** Evaluate the performance, security, and ethical considerations of AI applications through systematic audits.

   - **Benefits:** Identifies and rectifies any issues, ensuring ongoing compliance and reliability.

9. **Incorporation of Advanced Technologies:**

   - **Description:** Explore and integrate advancements in AI technologies.

   - **Applications:** Stay updated on the latest AI developments and incorporate relevant technologies to enhance capabilities.

   - **Benefits:** Positions construction projects at the forefront of technological innovation in quality control.

10. **Community Collaboration and Knowledge Sharing:**

- **Description:** Engage with the broader construction and AI communities for knowledge sharing.

- **Applications:** Participate in forums, conferences, and collaborative initiatives to share experiences and learn from industry peers.

- **Benefits:** Taps into collective expertise, fostering a culture of continuous improvement and innovation.

Continuous improvement in AI applications for quality control and assurance requires a proactive and collaborative approach. By embracing feedback, monitoring performance, and staying abreast of technological advancements, construction stakeholders can ensure that AI systems evolve to meet the changing demands of the industry while maintaining high standards of quality and safety.

## 9.6 Real World Example

1. **AI in Quality Control and Assurance:**

   **A) Automated Defect Detection (Company: Doxel):**

   - Doxel employs AI for quality control by automating defect detection in construction projects. The platform uses computer vision to analyze images and identify deviations from design plans, ensuring construction quality.

   **B) AI-driven Materials Testing (Company: Giatec):**

   - Giatec utilizes AI in materials testing for quality assurance. Their platform analyzes data from sensors embedded in concrete to assess material properties, ensuring compliance with quality standards in construction.

2. **Automated Inspection:**

   **C) Drone-based Inspection (Company: DJI):**

   - DJI, a leading drone manufacturer, facilitates automated inspection for quality assurance in construction. Drones equipped with cameras and AI technology conduct aerial inspections, identifying issues such as structural defects or construction deviations.

   **D) Robotic Inspections (Company: Boston Dynamics - Spot):**

   - Boston Dynamics' robot, Spot, is used for automated inspections in construction. Spot can navigate construction sites autonomously, capturing images and data for quality assurance purposes.

3. **Quality Assurance Algorithms:**

   **E) Computer Vision for Structural Monitoring (Company: NKSensors):**

   - NKSensors employs quality assurance algorithms based on computer vision for structural monitoring. Their technology analyzes visual data to detect anomalies and assess the structural integrity of civil infrastructure.

   **F) AI-powered QA Software (Company: Smartvid.io):**

- Smartvid.io provides AI-powered quality assurance software for construction. The platform utilizes machine learning algorithms to analyze images and videos, flagging potential issues and enhancing overall project quality.

4. **Non-Destructive Testing:**

**G) Ultrasonic Testing with AI (Company: Pundit AI - Proceq):**

- Proceq's Pundit AI incorporates non-destructive testing with AI for quality assurance. The platform uses ultrasonic testing data to assess concrete quality and detect potential defects, contributing to construction quality control.

**H) AI-enhanced Ground Penetrating Radar (Company: GPR Professional Services):**

- GPR Professional Services integrates AI with ground-penetrating radar (GPR) for non-destructive testing. The AI algorithms analyze GPR data to identify subsurface anomalies and assess the condition of buried structures in construction projects.

## 9.7 Chapter Summary: Key Points

1. AI-driven automated inspection in civil engineering employs computer vision for rapid defect detection, reducing reliance on manual inspections.

2. Machine learning predicts and prevents construction defects by analyzing historical data, enabling proactive measures and quality improvements.

3. AI-driven robotics enhance quality control by performing precise and repetitive tasks, reducing errors in construction projects.

4. Predictive analytics in quality control uses AI to analyze various data sources, predicting risks and variations for effective resource allocation.

5. Smart sensors and AI algorithms monitor real-time parameters in construction materials, enabling timely interventions and quality assurance.

6. Natural Language Processing (NLP) aids in analyzing textual data, ensuring construction projects adhere to relevant standards and regulations.

7. AI integration with digital twins simulates scenarios, assesses risks, and optimizes construction processes for enhanced quality assurance.

8. AI facilitates non-destructive testing methods, such as ultrasonic testing, thermal imaging, and acoustic monitoring, ensuring safety and durability.

9. Quality data management is streamlined with AI, enabling efficient storage, retrieval, and analysis for data-driven decision-making.

10. Quality assurance algorithms employ AI for defect detection, computer vision, pattern recognition, and materials compliance, ensuring adherence to standards.

11. AI-enhanced NDT techniques automate defect detection, analyze patterns, and provide real-time insights, improving accuracy and efficiency.

12. Certification and compliance frameworks ensure responsible and ethical AI use in construction, aligning with industry standards and regulations.

13. Continuous improvement in AI applications involves feedback loops, performance monitoring, user training, and collaboration with construction professionals.

14. Adaptive algorithms, benchmarking, and integration with BIM contribute to ongoing system audits, ensuring AI applications meet industry expectations.

15. Community collaboration and knowledge sharing foster a culture of continuous improvement, keeping construction projects at the forefront of AI-driven quality control innovation.

## 9.8 Concept Check: Q&A Sessions

1. How does AI revolutionize quality control and assurance processes in civil engineering and construction?

2. What is the significance of computer vision in quality control within construction?

3. How do AI algorithms contribute to defect detection and classification in construction materials and structures?

4. In what ways does AI-driven predictive maintenance minimize downtime and extend equipment lifespan in construction projects?

5. How do machine learning algorithms play a role in predicting and preventing construction defects?

6. What tasks do AI-driven robotics perform in quality control for construction projects?

7. How does AI contribute to real-time monitoring and assessment of construction projects?

8. In what ways does AI enhance data management for quality control in construction projects?

9. How does AI contribute to the development of digital twins in construction, and what benefits does it offer?

10. What role does AI play in non-destructive testing (NDT) techniques in civil engineering, and how does it enhance the overall quality control process?

# 10. Virtual Reality and Augmented Reality in Construction

Virtual Reality (VR) and Augmented Reality (AR) are transformative technologies that have found significant applications in the construction industry, reshaping the way projects are designed, visualized, and executed.

VR immerses users in a completely virtual environment, offering a 360-degree, computer-generated experience. In contrast, AR overlays digital information onto the real-world environment, enhancing the user's perception of the physical surroundings.

Virtual Reality (VR) and Augmented Reality (AR) are increasingly integral to the construction industry, offering diverse functionalities.

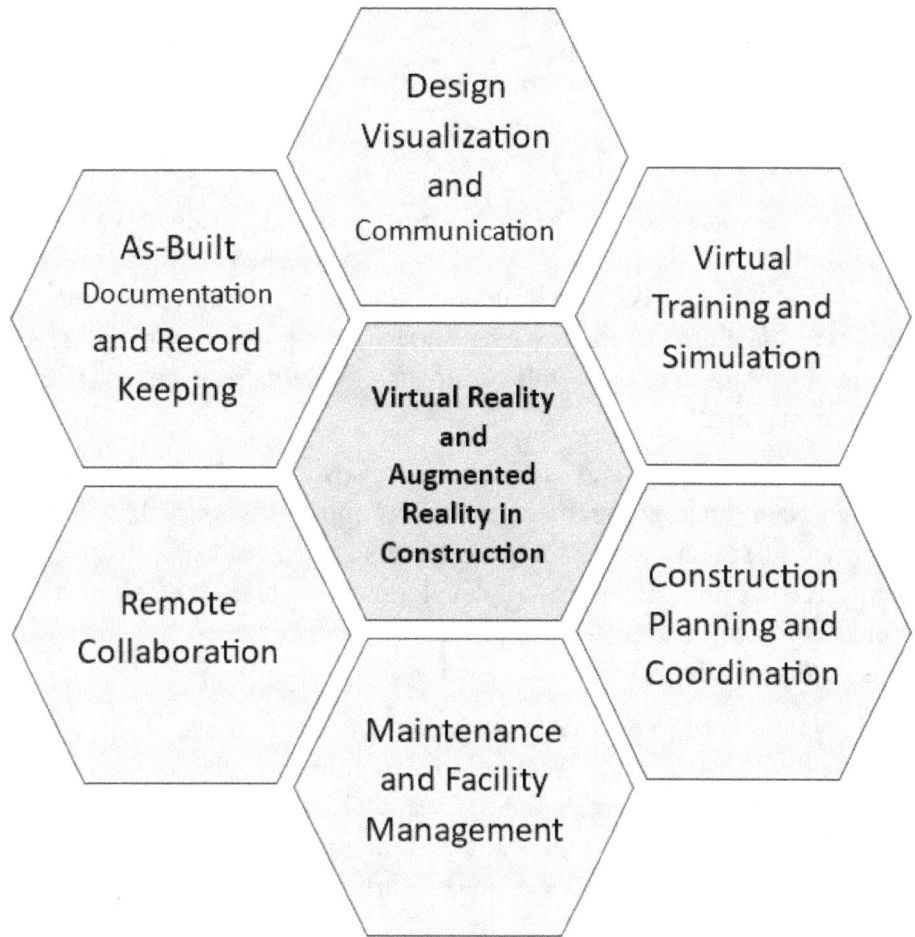

In terms of design, VR enables immersive experiences for architects and engineers within 3D models, fostering better communication. AR complements this by overlaying digital designs onto physical construction sites in real-time. Virtual training and simulation benefit construction workers through VR environments that simulate tasks and provide risk-free training, while AR offers on-site guidance with step-by-step visual instructions. Construction planning benefits from VR's visualization and simulation capabilities, aiding project managers in identifying issues and optimizing processes, while AR enhances coordination by overlaying construction plans onto physical sites. Maintenance and facility management benefit from VR-based training for maintenance personnel and AR applications providing real-time information to on-site workers.

VR facilitates remote collaboration through shared virtual environments for meetings, and AR supports remote assistance by enabling off-site experts to guide on-site workers. Finally, VR aids in virtual inspections of completed structures, while AR overlays digital documentation onto physical structures, simplifying access for maintenance teams. Together, these technologies significantly improve construction processes, from design and training to collaboration and ongoing maintenance.

One key application of VR in construction is immersive project visualization. Designers, architects, and stakeholders can use VR to explore virtual replicas of construction projects before they are built. This allows for a more comprehensive understanding of the spatial layout, design elements, and overall aesthetics, facilitating better-informed decision-making.

Construction training and simulation benefit significantly from VR. Workers can undergo realistic virtual training sessions that simulate construction scenarios, machinery operations, and safety protocols. This immersive training enhances skill development, reduces the risk of on-site accidents, and ensures that workers are well-prepared for real-world construction challenges.

AR is widely used in on-site construction activities. Construction workers equipped with AR devices, such as smart helmets or glasses, can overlay digital information onto their field of view. This includes real-time project data, blueprints, safety guidelines, and equipment instructions, improving efficiency and reducing errors during construction processes.

VR and AR contribute to the collaborative aspect of construction projects. Multiple stakeholders can participate in virtual meetings or walkthroughs, irrespective of geographical locations. This fosters effective communication, collaboration, and decision-making, leading to streamlined project management and reduced delays.

BIM (Building Information Modeling) is integrated with VR to create immersive 3D models that enhance project understanding. BIM models, when experienced in VR, provide a holistic view of the construction project, allowing stakeholders to identify potential issues, assess spatial relationships, and make informed design modifications before construction begins.

In construction planning and logistics, AR is utilized for visualizing complex construction sequences. Overlaying digital information onto the physical site enables project managers to assess the feasibility of construction sequences, track progress, and optimize resource allocation for more efficient project execution.

VR and AR technologies assist in facilities management post-construction. Maintenance teams can use AR to access digital overlays of equipment, identify issues, and receive step-by-step repair instructions. This improves the speed and accuracy of maintenance activities, reducing downtime and enhancing the lifespan of constructed facilities.

Safety training is a critical aspect of construction, and VR is increasingly used for simulating hazardous scenarios. Workers can undergo virtual safety drills, experiencing emergency situations in a controlled environment. This immersive training enhances safety awareness and preparedness for on-site challenges.

The adoption of VR and AR in construction aligns with the industry's pursuit of digitization and innovation. These technologies offer a dynamic, interactive, and efficient approach to project visualization, training, collaboration, and maintenance, ultimately contributing to improved construction processes, enhanced safety standards, and more successful project outcomes.

## 10.1 Immersive Technologies

Immersive technologies, including Virtual Reality (VR) and Augmented Reality (AR), have been increasingly integrated into the construction industry, offering innovative solutions for design, visualization, and project management.

Here are insights into the application of immersive technologies in construction:

1. **Virtual Reality (VR) in Construction:**

   - **Description:** VR creates a fully immersive, computer-generated environment that users can explore and interact with.

   - **Applications:**

- o **Design Visualization:** Architects and engineers can use VR to visualize and navigate through complex 3D models, gaining a better understanding of spatial relationships.

- o **Virtual Walkthroughs:** Stakeholders can take virtual tours of construction sites before physical construction begins, helping identify potential issues.

- o **Training and Simulation:** VR is employed for training construction workers in a safe and controlled virtual environment, simulating real-world scenarios.

2. **Augmented Reality (AR) in Construction:**

   - • **Description:** AR overlays digital information onto the user's view of the real world, enhancing their perception.

   - • **Applications:**

     - o **On-Site Visualization:** Construction professionals use AR to overlay digital models onto physical structures, aiding in on-site decision-making.

     - o **Maintenance and Repairs:** AR facilitates the overlay of relevant information, such as maintenance instructions or equipment details, onto physical assets.

     - o **Safety Enhancements**: AR can provide real-time safety information, such as highlighting potential hazards on construction sites.

3. **Digital Twin Technology:**

   - • **Description:** Digital twins are virtual replicas of physical objects or systems, often used in combination with VR and AR.

   - • **Applications:**

     - o **Real-Time Monitoring:** Digital twins enable real-time monitoring of construction projects, allowing stakeholders to track progress and identify issues.

     - o **Predictive Analytics:** By combining data from sensors with digital twins, construction professionals can use predictive analytics for better decision-making.

4. **Immersive Project Collaboration:**

- **Description:** VR and AR facilitate collaborative experiences among project stakeholders.

- **Applications:**

  o **Remote Collaboration:** VR enables geographically dispersed teams to collaborate in a shared virtual space, fostering effective communication.

  o **Design Reviews:** Stakeholders can participate in virtual design reviews, providing feedback on models and making collaborative decisions.

5. **Construction Site Planning and Simulation:**

- **Description:** VR is utilized for simulating construction site scenarios and planning.

- **Applications:**

  o **Logistical Planning:** VR helps in planning construction logistics, optimizing material flow, and minimizing on-site congestion.

  o **Safety Training:** VR-based safety simulations allow workers to experience and learn from potential on-site hazards in a controlled environment.

6. **Client and Stakeholder Engagement:**

- **Description:** Immersive technologies enhance client engagement and communication.

- **Applications:**

  o **Virtual Showcases:** VR is used to create immersive presentations and showcases for clients, providing a realistic preview of the final project.

  o **Interactive Site Tours:** AR-enhanced site tours offer stakeholders interactive and informative experiences during project walkthroughs.

7. **Integration with Building Information Modeling (BIM):**

- **Description:** VR and AR technologies are integrated with BIM for enhanced project visualization.

- **Applications:**

  o **BIM Visualization:** VR and AR enable stakeholders to interact with BIM models in a more immersive and intuitive way.

  o **Construction Planning:** Construction teams can use AR to overlay BIM data onto the physical construction site, improving accuracy during the building process.

8. **Progress Monitoring and Documentation:**

- **Description:** Immersive technologies assist in monitoring construction progress and documenting changes.

- **Applications:**

  o **Progress Tracking:** VR and AR provide visual documentation of construction phases, aiding in progress tracking and quality control.

  o **As-Built vs. As-Planned Comparisons:** Construction professionals use AR to compare the as-built state with the initial plans, identifying any deviations.

9. **Training Simulators for Heavy Equipment Operation:**

- **Description:** VR is employed to create realistic training simulators for operating heavy construction equipment.

- **Applications:** Trainees can practice operating equipment in a virtual environment before working on actual construction sites, enhancing safety and skill development.

10. **Visualization for Stakeholder Communication:**

- **Description:** Immersive technologies enhance communication with various stakeholders.

- **Applications:**

  o **Public Engagement:** VR and AR can be used for public engagement by providing immersive experiences, such as virtual open houses or community presentations.

- o **Investor Presentations:** Construction companies use immersive technologies to create compelling presentations for potential investors, showcasing project details in a visually engaging manner.

The integration of VR and AR in construction represents a paradigm shift in project design, communication, and collaboration, offering substantial benefits in terms of efficiency, safety, and stakeholder engagement. As technology continues to advance, the applications of immersive technologies in construction are likely to expand, further transforming the industry's practices and processes.

## 10.2 Virtual Design and Construction

Virtual Design and Construction (VDC) is a comprehensive approach that leverages Virtual Reality (VR) and Augmented Reality (AR) technologies in the construction industry.

It integrates digital modeling, simulation, and collaboration to enhance the design, planning, and execution of construction projects. Here's an overview of how VDC utilizes VR and AR in construction:

1. **Building Information Modeling (BIM) Integration:**

   - **Description:** VDC begins with the creation of detailed 3D digital models using BIM software, incorporating information about the project's physical and functional characteristics.

   - **VR Application:** VR is used to immerse stakeholders in a virtual environment where they can navigate and explore the

BIM models, gaining a better understanding of the project's design and functionality.

- **AR Application:** AR overlays BIM data onto the physical construction site, providing on-site workers with real-time information about the project elements.

2. **Design Visualization:**

- **Description:** VDC employs VR and AR to visualize complex designs and facilitate better communication among project stakeholders.

- **VR Application:** VR allows designers and stakeholders to experience a virtual walkthrough of the project, exploring the design from various angles and perspectives.

- **AR Application:** AR enhances on-site visualization by overlaying digital design information onto the physical environment, aiding construction teams in understanding design specifications.

3. **Clash Detection and Coordination:**

- **Description:** VDC uses VR and AR to identify and resolve clashes or conflicts in the design phase before physical construction begins.

- **VR Application:** VR enables immersive clash detection, where stakeholders can identify and address clashes between various building components in a virtual space.

- **AR Application:** AR assists on-site workers by highlighting clashes or coordination issues on the physical construction site, improving accuracy during the building process.

4. **Construction Planning and Simulation:**

- **Description:** VDC utilizes VR for simulating construction sequences and logistics, optimizing project planning.

- **VR Application:** VR allows construction teams to virtually simulate the construction process, optimizing resource allocation, and identifying potential challenges.

- **AR Application:** AR supports on-site construction planning by overlaying digital construction sequences onto the physical environment, aiding workers in executing tasks efficiently.

5. **Collaborative Design Review:**

    - **Description:** VDC fosters collaboration among project stakeholders by enabling them to review and provide feedback on the design.

    - **VR Application:** VR facilitates collaborative design reviews in a virtual environment, allowing stakeholders from different locations to participate and contribute.

    - **AR Application:** AR enhances on-site collaboration by overlaying design-related information onto the physical environment, enabling real-time discussions among construction teams.

6. **As-Built vs. As-Planned Comparisons:**

    - **Description:** VDC uses VR and AR to compare the as-built state of the construction project with the initial design plans.

    - **VR Application:** VR allows stakeholders to visually compare the as-built construction with the original design in a virtual space.

    - **AR Application:** AR enables on-site workers to overlay design plans onto the physical construction site, facilitating accurate as-built comparisons.

7. **Training and Safety Simulations:**

    - **Description:** VDC incorporates VR for training simulations to enhance worker skills and improve safety awareness.

    - **VR Application:** VR-based safety simulations allow workers to experience and learn from potential on-site hazards in a controlled virtual environment.

    - **AR Application:** AR enhances on-site safety by providing real-time safety information, such as hazard warnings and safety protocols.

8. **Project Documentation and Reporting:**

- **Description:** VDC uses VR and AR to document project progress and generate reports.

- **VR Application:** VR captures visual documentation of construction phases for reporting and analysis.

- **AR Application:** AR assists in real-time data analysis and reporting on the physical construction site, providing insights into project performance.

9. **Remote Collaboration:**

- **Description:** VDC enables remote collaboration among project teams through VR and AR technologies.

- **VR Application:** VR facilitates virtual meetings and collaborative sessions for stakeholders located in different geographical locations.

- **AR Application:** AR enhances remote collaboration by overlaying digital information onto the physical construction site, allowing remote experts to provide guidance.

10. **Public Engagement:**

- **Description:** VDC utilizes VR and AR for engaging the public and stakeholders in the project.

- **VR Application:** VR creates immersive presentations and showcases for public engagement, providing a realistic preview of the final project.

- **AR Application:** AR-enhanced site tours offer stakeholders interactive and informative experiences during project walkthroughs.

Virtual Design and Construction, powered by VR and AR technologies, significantly enhances the construction process by improving communication, collaboration, and decision-making throughout the project lifecycle. This integrated approach contributes to more efficient and accurate construction practices.

# 10.3 AR for On-site Navigation

Augmented Reality (AR) plays a crucial role in on-site navigation in the construction industry, enhancing efficiency, accuracy, and safety.

Here's how AR is utilized for on-site navigation in conjunction with Virtual Reality (VR) in construction:

1. **Digital Information Overlay:**

    - **Description:** AR overlays digital information, such as construction plans, annotations, and navigational cues, onto the physical construction site in real-time.

    - **Functionality:** On-site workers wearing AR devices, such as smart glasses or AR-enabled helmets, can see relevant information overlaid on their field of view as they navigate the construction site.

2. **Virtual Wayfinding:**

   - **Description:** AR assists on-site workers in wayfinding by providing virtual navigation paths and directions superimposed on the physical environment.

   - **Functionality:** On-site navigation instructions, including arrows, symbols, or virtual markers, guide workers to specific locations, reducing the likelihood of getting lost.

3. **Site Element Identification:**

   - **Description:** AR enables workers to identify and understand various site elements, such as structural components, utilities, or safety zones.

   - **Functionality:** On-site workers can use AR to visualize information about specific elements by pointing their devices at them, gaining insights into the construction layout.

4. **Real-time Construction Progress Monitoring:**

   - **Description:** AR provides real-time updates on construction progress, allowing workers to visualize completed and ongoing tasks.

   - **Functionality:** On-site workers can access augmented views that highlight completed construction elements or display the progress of ongoing activities, aiding in project tracking.

5. **Safety Hazard Alerts:**

   - **Description:** AR enhances on-site safety by providing real-time alerts and warnings about potential hazards.

   - **Functionality:** Workers can receive visual cues in their AR devices, warning them about safety hazards, restricted areas, or the presence of heavy equipment, promoting a safer working environment.

6. **Equipment Navigation Assistance:**

- **Description:** AR assists equipment operators in navigating construction machinery within the site.

- **Functionality:** AR overlays information on construction equipment, offering operators guidance on optimal routes, potential obstacles, and safety considerations while navigating the site.

7. **Coordination with BIM Models:**

- **Description:** AR integrates with Building Information Modeling (BIM) data, allowing workers to align virtual models with the physical construction site.

- **Functionality:** On-site workers can use AR to compare BIM models with the actual construction, ensuring alignment and accuracy during the building process.

8. **Interactive Project Communication:**

- **Description:** AR facilitates interactive communication among on-site teams, allowing them to share insights and coordinate activities.

- **Functionality:** On-site workers wearing AR devices can participate in virtual meetings, share annotations, and collaborate on project-related tasks in real-time.

9. **Dynamic Design Visualization:**

- **Description:** AR enables dynamic visualization of design changes or updates in the construction process.

- **Functionality:** On-site workers can see updated design plans, modifications, or additional information overlaid on the construction site, ensuring they work with the latest data.

10. **Training and Onboarding:**

- **Description:** AR aids in on-site training and onboarding processes for new workers.

- **Functionality:** New workers can use AR to access training modules, overlaying instructional information and guidance on their real-time view, facilitating a quicker learning curve.

The integration of AR for on-site navigation in construction not only improves efficiency and accuracy but also contributes to a safer working environment. By providing real-time information and visual overlays, AR enhances the overall navigation experience for on-site workers, promoting better collaboration and informed decision-making.

# 10.4 Training and Simulation

Training and simulation using Virtual Reality (VR) and Augmented Reality (AR) in construction offer innovative and effective approaches to enhance learning, improve safety, and streamline project workflows.

Here's an overview of how VR and AR are utilized for training and simulation in the construction industry:

**A. Virtual Reality (VR) in Training and Simulation:**

   **1. Immersive Job Site Familiarization:**

- **Description:** VR allows trainees to explore virtual job sites, becoming familiar with different project phases, layouts, and potential hazards.

- **Benefits:** Enhances job site orientation, reducing the learning curve for new workers and improving overall job site awareness.

2. **Equipment Operation Training:**

   - **Description:** VR simulations provide hands-on training for operating construction machinery and equipment in a virtual environment.

   - **Benefits:** Improves equipment operator skills, reduces the risk of on-site accidents, and allows for repetitive practice in a controlled setting.

3. **Safety Training and Emergency Response:**

   - **Description:** VR scenarios simulate various safety hazards and emergency situations, allowing workers to practice responses.

   - **Benefits:** Enhances safety awareness, ensures workers are prepared for emergencies, and reduces the likelihood of accidents through immersive training.

4. **Construction Process Visualization:**

   - **Description:** VR enables trainees to visualize entire construction processes, from project initiation to completion, in a 3D virtual environment.

   - **Benefits:** Improves understanding of project workflows, sequencing, and collaboration, facilitating better-informed decision-making.

5. **Collaborative Design Reviews:**

   - **Description:** VR facilitates collaborative design reviews, allowing stakeholders to interact with 3D models and provide feedback in real-time.

   - **Benefits:** Enhances communication among project teams, accelerates design iterations, and ensures alignment on project specifications.

6. **Site Logistics Planning:**

- **Description:** VR can simulate site logistics and construction phasing, helping planners optimize material flow, storage, and equipment placement.

- **Benefits:** Improves logistical efficiency, minimizes on-site congestion, and enhances overall project planning.

## A. Augmented Reality (AR) in Training and Simulation:

### 1. On-site Work Instructions:

- **Description:** AR overlays step-by-step work instructions onto physical construction elements, guiding workers through tasks.

- **Benefits:** Improves task accuracy, reduces errors, and provides real-time guidance for on-site workers.

### 2. Equipment Maintenance Training:

- **Description:** AR provides maintenance instructions and visualizations overlaid on construction equipment, facilitating hands-on training.

- **Benefits:** Enhances equipment maintenance skills, reduces downtime, and ensures timely and accurate repairs.

### 3. Interactive Equipment Operation Guides:

- **Description:** AR-enhanced operation manuals and guides are superimposed on construction equipment, offering interactive training experiences.

- **Benefits:** Improves equipment operation skills, simplifies training processes, and reduces the need for extensive paper manuals.

### 4. As-Built Visualization:

- **Description:** AR overlays as-built or updated design information onto physical structures, allowing workers to see real-time changes.

- **Benefits:** Ensures workers are aware of design modifications, reducing errors and streamlining construction processes.

5. **Real-time Collaboration and Communication:**

- **Description:** AR enables real-time collaboration by allowing multiple users to view and interact with construction models simultaneously.

- **Benefits:** Facilitates collaborative decision-making, reduces project delays, and enhances communication among team members.

6. **Quality Control and Inspection:**

- **Description:** AR assists in quality control by overlaying inspection criteria onto physical structures, guiding inspectors during on-site assessments.

- **Benefits:** Improves inspection accuracy, accelerates the inspection process, and ensures compliance with quality standards.

Both VR and AR contribute to a more immersive, interactive, and effective training environment in construction. These technologies not only enhance skills but also promote safety, collaboration, and efficiency throughout the construction project lifecycle.

## 10.5 Integrating VR/AR with AI

Integrating Virtual Reality (VR) and Augmented Reality (AR) with Artificial Intelligence (AI) in construction brings about a synergistic combination that enhances project planning, design, construction, and maintenance.

Here's how the integration of VR/AR with AI can benefit the construction industry:

### A. Integration of AI with Virtual Reality (VR) in Construction:

1. **Generative Design and VR:**

   - **Description:** AI-driven generative design algorithms can create numerous design variations based on project requirements. VR enables stakeholders to explore and interact with these designs in a 3D virtual environment.

- **Benefits:** Accelerates the design exploration process, improves collaboration, and allows for real-time design modifications.

2. **AI-Enhanced Construction Simulation:**

   - **Description:** AI algorithms can predict and simulate construction processes, optimizing schedules, resource allocation, and logistics. VR provides a platform to visualize and interact with these simulations.

   - **Benefits:** Enhances decision-making by providing stakeholders with a realistic view of project timelines, potential bottlenecks, and resource requirements.

3. **Realistic Job Site Visualization with AI:**

   - **Description:** AI algorithms analyze project data to generate realistic virtual job site environments. VR allows users to immerse themselves in these environments for better project understanding.

   - **Benefits:** Improves job site familiarization, enhances safety training, and facilitates collaboration by providing a visual representation of the construction site.

4. **AI-Driven Predictive Maintenance in VR:**

   - **Description:** AI algorithms predict equipment maintenance needs based on historical data. VR enables maintenance personnel to virtually inspect and plan repairs.

   - **Benefits:** Reduces equipment downtime, improves maintenance efficiency, and extends the lifespan of construction machinery.

5. **Collaborative Design Review with AI Insights:**

   - **Description:** AI can provide insights and recommendations during collaborative design reviews. VR allows stakeholders to interact with AI-generated insights in real-time.

- **Benefits:** Enhances design quality, accelerates design iterations, and ensures that AI-driven recommendations are considered in the collaborative decision-making process.

## B. Integration of AI with Augmented Reality (AR) in Construction:

### 1. AI-Enhanced Real-time Data Overlays:

- **Description:** AI algorithms process real-time project data, and AR overlays this information onto physical structures, providing on-site workers with contextual insights.

- **Benefits:** Improves on-site decision-making, enhances situational awareness, and ensures that workers have access to up-to-date information.

### 2. AI-Driven Equipment Operation Guidance in AR:

- **Description:** AI algorithms can analyze equipment performance data and provide real-time guidance to equipment operators through AR overlays.

- **Benefits:** Enhances equipment operation skills, reduces errors, and ensures optimal performance through AI-driven guidance.

### 3. AR-Enhanced Quality Control with AI Analysis:

- **Description:** AI algorithms analyze construction elements for quality. AR overlays inspection criteria onto physical structures, guiding inspectors during on-site quality control.

- **Benefits:** Improves inspection accuracy, accelerates the inspection process, and ensures compliance with quality standards.

### 4. AI-Integrated Safety Training in AR:

- **Description:** AI algorithms analyze safety data, and AR overlays safety instructions and hazard alerts onto physical job site elements.

- **Benefits:** Enhances safety training, reduces the risk of accidents, and ensures that workers are aware of potential hazards in real-time.

5. **AI-Enhanced As-Built Visualization in AR:**

   - **Description:** AI processes design changes, and AR overlays updated design information onto physical structures, allowing workers to see real-time changes.

   - **Benefits:** Reduces errors associated with outdated design information, streamlines construction processes, and ensures alignment with the latest design modifications.

C. **Overall Benefits of VR/AR and AI Integration in Construction:**

1. **Improved Decision-Making:**

   - VR and AR provide immersive environments for stakeholders to make informed decisions based on AI-driven insights.

2. **Enhanced Collaboration:**

   - Collaborative design reviews and real-time data sharing in VR/AR foster better collaboration among project teams, with AI-driven insights enhancing the decision-making process.

3. **Efficient Training:**

   - VR and AR simulations, enriched by AI insights, provide realistic training scenarios for equipment operation, safety procedures, and job site tasks.

4. **Optimized Construction Processes:**

   - AI-driven simulations in VR optimize construction processes, schedules, and resource allocation, leading to more efficient project outcomes.

5. **Improved Safety:**

   - AR overlays safety information in real-time, guided by AI insights, enhancing on-site safety training and reducing the risk of accidents.

6. **Streamlined Maintenance:**

- AI-driven predictive maintenance, visualized through VR and AR, improves equipment maintenance planning and reduces downtime.

7. **Quality Assurance:**

- AR-enhanced quality control, guided by AI analysis, ensures that construction elements meet quality standards, reducing defects and rework.

The integration of VR/AR with AI holds the potential to revolutionize the construction industry by creating smarter, more efficient, and safer project environments.

## 10.6 Real World Examples

1. **Virtual Reality and Augmented Reality:**

   - **Virtual Project Collaboration (Company: IrisVR):**
     - IrisVR offers virtual reality solutions for civil engineering and construction, enabling project collaborators to experience 3D models in immersive environments. This enhances design communication and collaboration among stakeholders.

   - **AR for Construction Visualization (Company: Trimble - SiteVision):**
     - Trimble's SiteVision utilizes augmented reality for on-site construction visualization. The AR application overlays digital design models onto the physical construction site, providing real-time visualizations for improved alignment with design plans.

2. **Immersive Technologies:**

   - **Immersive 3D Modeling (Company: Unity Reflect):**
     - Unity Reflect provides immersive 3D modeling solutions for civil engineering and construction. The platform enables users to create and experience interactive 3D models in real-time, facilitating design exploration and collaboration.

   - **VR Safety Training (Company: Mace Group):**
     - Mace Group uses virtual reality for immersive safety training in construction. VR simulations recreate on-site scenarios, allowing workers to undergo safety training in a controlled, realistic environment.

3. **Virtual Design and Construction:**

   - **VR-based Design Review (Company: Autodesk BIM 360):**
     - Autodesk BIM 360 integrates virtual reality for design review in construction. The platform enables stakeholders

to conduct immersive design reviews, enhancing the understanding and evaluation of construction projects.

- **VR for Construction Sequencing (Company: Theia Interactive - Clarity):**

  o Clarity, developed by Theia Interactive, utilizes virtual reality for construction sequencing. The VR platform allows project teams to visualize and simulate construction sequences, aiding in project planning and coordination.

4. **AR for On-site Navigation:**

- **AR Navigation Helmets (Company: DAQRI):**

  o DAQRI develops augmented reality navigation helmets for on-site use in construction. These helmets overlay digital information onto the physical environment, providing on-site workers with real-time guidance and data.

- **AR-enabled Safety Glasses (Company: Microsoft - HoloLens):**

  o Microsoft's HoloLens is used in construction with AR-enabled safety glasses. These glasses provide on-site workers with augmented information, such as design overlays and safety instructions, enhancing on-site navigation and tasks.

5. **Training and Simulation using VR and AR:**

- **VR Crane Operation Simulation (Company: Serious Labs):**

  o Serious Labs offers virtual reality simulations for crane operation training in construction. VR technology provides a realistic and safe environment for operators to practice and improve their skills.

- **AR On-site Equipment Guidance (Company: Trimble - Connect for HoloLens):**

  o Trimble's Connect for HoloLens uses augmented reality for on-site equipment guidance. The AR application provides on-site workers with visual instructions and information overlays, guiding them through tasks and procedures in real-time.

6. **Integrating VR/AR with AI:**

- **AI-driven VR Design Optimization (Company: NVIDIA - Omniverse):**

  o NVIDIA Omniverse integrates artificial intelligence with virtual reality for design optimization in construction. AI algorithms assist in generative design, optimizing designs for efficiency and aesthetics within the VR environment.

- **AI-enhanced AR Construction Coordination (Company: Miralupa):**

  o Miralupa integrates AI with augmented reality for construction coordination. The platform uses AI algorithms to analyze construction plans and data, providing on-site workers with AR overlays for improved coordination and reduced errors.

## 10.7 Chapter Summary: Key Points

1. Virtual Reality (VR) and Augmented Reality (AR) are transforming construction by offering immersive experiences for design visualization and real-time on-site guidance.

2. VR is extensively used for immersive project visualization, allowing stakeholders to explore virtual replicas of construction projects before physical construction.

3. Construction training benefits from VR with realistic virtual sessions simulating construction scenarios, machinery operations, and safety protocols.

4. Augmented Reality (AR) enhances on-site construction activities by overlaying digital information onto the real-world environment, improving efficiency and reducing errors.

5. VR and AR contribute to collaborative construction projects, enabling virtual meetings, walkthroughs, and design reviews for geographically dispersed teams.

6. Building Information Modeling (BIM) is integrated with VR to create immersive 3D models, aiding project understanding and issue identification.

7. AR is utilized in construction planning, visualizing complex construction sequences and optimizing resource allocation for efficient project execution.

8. VR and AR assist in facilities management post-construction, providing maintenance teams with digital overlays and real-time information for streamlined activities.

9. VR is employed for safety training, simulating hazardous scenarios and enhancing workers' safety awareness in a controlled environment.

10. VR and AR integration in construction enhances client engagement with virtual showcases, interactive site tours, and immersive presentations.

11. Virtual Design and Construction (VDC) leverages VR and AR for detailed 3D modeling, design visualization, clash detection, and collaborative design reviews.

12. VDC utilizes AR for on-site visualization, maintenance, and real-time collaboration, improving overall efficiency and accuracy in construction projects.

13. AR plays a crucial role in on-site navigation, overlaying digital information, providing wayfinding, and enhancing site element identification for workers.

14. VR and AR offer innovative approaches to training and simulation in construction, improving job site familiarization, equipment operation, safety training, and collaborative design reviews.

15. Integrating VR/AR with AI in construction enhances decision-making, collaboration, training, construction processes, safety, maintenance, and quality assurance, revolutionizing the industry.

## 10.8 Concept Check: Q & A Sessions

1.  What is the primary difference between Virtual Reality (VR) and Augmented Reality (AR) in the context of the construction industry?

2.  How does Virtual Reality (VR) contribute to design processes in construction, and what benefits does it offer architects and engineers?

3.  What is one key application of Augmented Reality (AR) in on-site construction activities, and how does it improve efficiency?

4.  How does Virtual Reality (VR) contribute to safety training in the construction industry, and what are the benefits for workers?

5.  In what way does Augmented Reality (AR) enhance facilities management post-construction, and what advantages does it provide for maintenance teams?

6.  How does Virtual Reality (VR) contribute to remote collaboration in the construction industry, and what is the key feature that facilitates this?

7.  How does Augmented Reality (AR) benefit construction planning and logistics, and what role does it play in optimizing processes?

8.  What is the significance of integrating Building Information Modeling (BIM) with Virtual Reality (VR) in construction, and how does it enhance project visualization?

9.  How does Virtual Design and Construction (VDC) leverage Virtual Reality (VR) and Augmented Reality (AR) technologies, and what benefits does it bring to project planning?

10. How does Augmented Reality (AR) play a crucial role in on-site navigation in the construction industry, and what information does it overlay onto the physical construction site?

# 11. AI in Energy Efficiency and Sustainability

Artificial Intelligence (AI) is playing a pivotal role in advancing energy efficiency and sustainability initiatives across various sectors.

One key application of AI is in optimizing energy consumption in buildings. Smart building systems equipped with AI algorithms can analyze data from sensors and control devices to adjust lighting, heating, and cooling in real-time, ensuring efficient energy use while maintaining occupant comfort.

In the realm of renewable energy, AI contributes to enhancing the efficiency of solar and wind power systems. AI algorithms can predict energy production patterns, optimize the operation of renewable energy sources, and integrate them into the existing power grid. This aids in reducing reliance on non-renewable energy sources and promoting a more sustainable energy mix.

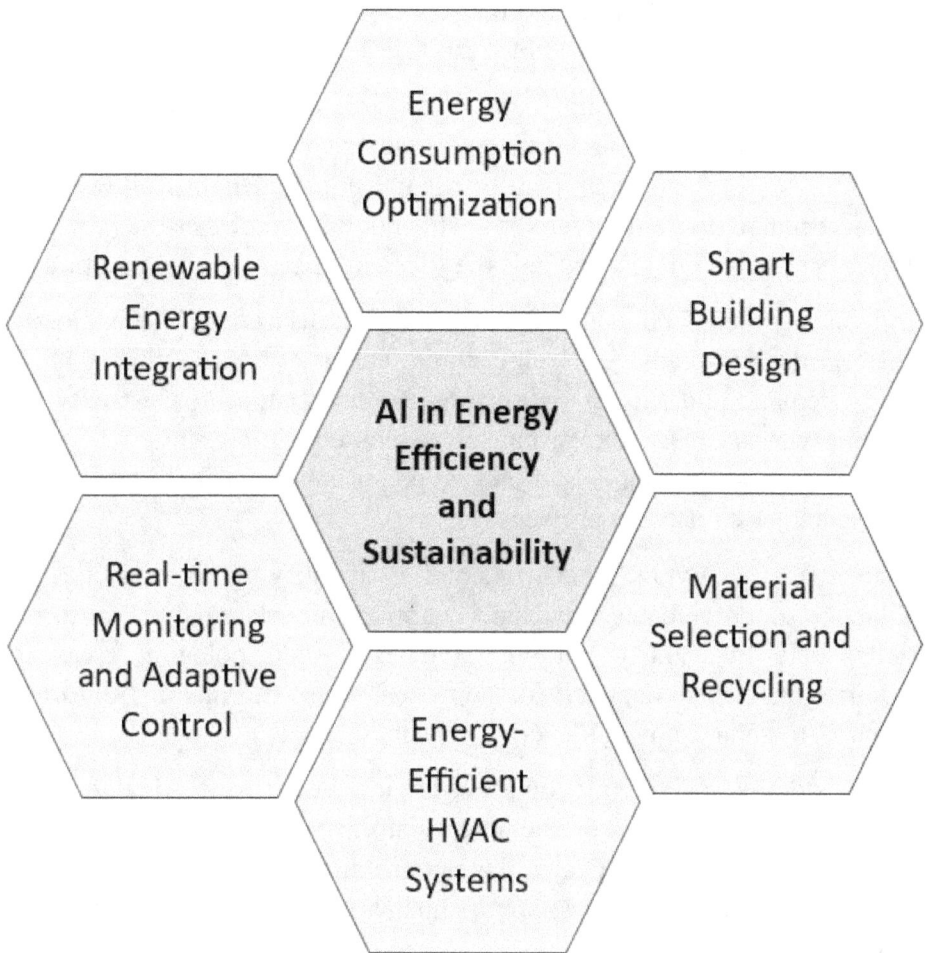

Artificial Intelligence (AI) plays a central role in advancing energy efficiency and sustainability in civil engineering and construction by optimizing energy consumption, enhancing smart building design through generative tools, and facilitating material selection with environmental considerations. AI contributes to construction waste management by identifying opportunities for recycling and reuse. It ensures energy-efficient HVAC systems through predictive maintenance, extending equipment lifespan. Real-time monitoring and adaptive controls, enabled by AI and IoT integration, optimize energy usage in buildings. Moreover, AI aids in the integration of renewable energy sources into the grid, optimizing energy production, managing fluctuations, and enhancing overall energy resilience. Collectively, these AI-driven functions support the adoption of more sustainable and environmentally friendly practices in the construction industry, aligning with global efforts to reduce environmental impact and promote responsible resource use.

AI-driven predictive maintenance is crucial for optimizing the performance of energy infrastructure. Machine learning models analyze data from sensors and equipment to predict potential faults, allowing for proactive maintenance interventions. This not only improves the reliability of energy systems but also extends the lifespan of critical infrastructure.

Grid management and demand response benefit from AI technologies. Smart grids, powered by AI algorithms, can analyze real-time data on energy demand and supply, optimizing the distribution of electricity and reducing wastage. Demand response systems enable consumers to adjust their energy consumption based on grid conditions, contributing to a more flexible and sustainable energy ecosystem.

AI in energy analytics is instrumental in identifying opportunities for efficiency improvements. Machine learning models analyze historical energy data to identify patterns, anomalies, and potential areas for optimization. This data-driven approach aids in making informed decisions to enhance overall energy efficiency.

Energy storage systems are optimized with the help of AI algorithms. Machine learning models predict energy storage needs, enabling efficient charging and discharging cycles. This contributes to balancing supply and demand, reducing grid stress, and enhancing the overall reliability of energy storage solutions.

In the transportation sector, AI plays a role in promoting sustainable practices. Smart traffic management systems, powered by AI, optimize traffic flow, reduce congestion, and minimize fuel consumption. Additionally, AI algorithms are employed in electric vehicle (EV) charging infrastructure to optimize charging schedules and locations.

AI supports the development of intelligent energy management systems for industries. These systems analyze production processes, energy consumption patterns, and environmental factors to optimize energy use, reduce waste, and minimize the carbon footprint of industrial operations.

Environmental monitoring and conservation efforts benefit from AI applications. AI algorithms analyze data from sensors and satellite imagery to monitor ecosystems, track environmental changes, and assess the impact of human activities. This information aids in making informed decisions for sustainable resource management.

The integration of AI in energy efficiency and sustainability is a dynamic and evolving field. As technology continues to advance, AI's role in optimizing energy use, promoting renewable sources, and fostering sustainable practices is expected to expand, contributing to a more resilient and environmentally conscious energy landscape.

# 11.1 Energy Modeling

Energy modeling plays a crucial role in enhancing energy efficiency and sustainability in the context of AI applications in civil engineering and construction.

Here are key aspects and considerations related to energy modeling for AI in this domain:

1. **Building Energy Modeling (BEM):**

   - **Purpose:** BEM is essential for assessing and optimizing the energy performance of buildings.

   - **AI Integration:** AI algorithms can enhance BEM by learning from historical data, weather patterns, and building usage to predict and optimize energy consumption.

2. **Predictive Maintenance with AI:**

- **Purpose:** AI can predict equipment failures and inefficiencies, enabling proactive maintenance.

- **Energy Modeling Integration:** Incorporate predictive maintenance data into energy models to optimize the performance of HVAC systems and other building components.

3. **Occupancy and Usage Pattern Analysis:**

- **Purpose:** Understanding how spaces are used helps optimize heating, cooling, and lighting systems.

- **AI Integration:** Machine learning models can analyze occupancy patterns and adapt building systems in real-time, reducing unnecessary energy consumption.

4. **Smart Grids and Energy Management Systems:**

- **Purpose:** Enhance the efficiency of energy distribution and consumption in a larger urban context.

- **AI Integration:** AI algorithms can optimize energy flow, predict demand, and dynamically adjust energy distribution in response to changing conditions.

5. **Life Cycle Assessment (LCA):**

- **Purpose:** Assessing the environmental impact of construction materials and processes over the entire life cycle.

- **AI Integration:** Use AI for data analysis and optimization strategies to minimize the environmental footprint of construction projects.

6. **Material Selection and Optimization:**

- **Purpose:** Choose materials with lower embodied energy and environmental impact.

- **AI Integration:** Machine learning can analyze material properties, costs, and environmental impact, aiding in the selection of sustainable options.

7. **Renewable Energy Integration:**

- **Purpose:** Increase the use of renewable energy sources in buildings and construction projects.

- **AI Integration:** AI algorithms can predict renewable energy availability, optimize energy storage, and manage the integration of renewable sources into the grid.

8. **Regulatory Compliance and Certification:**

- **Purpose:** Ensure that construction projects comply with energy efficiency standards and certifications.

- **AI Integration:** AI can assist in monitoring and ensuring ongoing compliance by analyzing real-time data and suggesting adjustments to meet standards.

9. **Continuous Monitoring and Feedback:**

- **Purpose:** Regularly monitor and provide feedback on the energy performance of buildings.

- **AI Integration:** Implement AI-powered monitoring systems that continuously analyze data and provide recommendations for improving energy efficiency.

10. **Collaborative Design and Decision-Making:**

- **Purpose:** Foster collaboration among architects, engineers, and stakeholders to make informed decisions.

- **AI Integration:** AI can facilitate collaborative design by analyzing diverse datasets and providing insights that contribute to energy-efficient and sustainable design decisions.

In summary, integrating AI with energy modeling in civil engineering and construction can lead to more efficient and sustainable practices by optimizing building performance, predicting maintenance needs, and promoting the use of renewable energy sources. This holistic approach considers the entire life cycle of construction projects, from design to operation and maintenance.

## 11.2 Green Building Practices

Green building practices, combined with AI applications, can significantly contribute to energy efficiency and sustainability in the construction industry.

Here's how AI can be integrated into various aspects of green building practices:

1.  **Site Selection and Planning:**

    - **AI Integration:** Use AI algorithms to analyze data related to climate, topography, and environmental impact to optimize site selection and design. This ensures that buildings are positioned to take advantage of natural resources such as sunlight and wind for energy efficiency.

2. **Energy-Efficient Design:**

   - **AI Integration:** Employ AI in the design phase to optimize building layouts, window placements, and material selection for maximum energy efficiency. AI can analyze vast datasets and simulate different design scenarios to identify the most sustainable options.

3. **Smart HVAC Systems:**

   - **Green Practice:** Use energy-efficient HVAC systems and passive design strategies.

   - **AI Integration:** AI can optimize HVAC performance by learning from historical data, predicting usage patterns, and adjusting heating and cooling systems in real-time based on occupancy and environmental conditions.

4. **Lighting Optimization:**

   - **Green Practice:** Implement energy-efficient lighting systems and controls.

   - **AI Integration:** AI can analyze patterns of natural light, occupancy, and user preferences to optimize artificial lighting. Smart lighting systems can adjust brightness and color temperature dynamically for energy savings.

5. **Water Efficiency:**

   - **Green Practice:** Incorporate water-efficient fixtures and landscaping designs.

   - **AI Integration:** Use AI to predict water usage patterns, optimize irrigation systems, and detect and address water leaks promptly, ensuring efficient water management.

6. **Material Selection:**

   - **Green Practice:** Choose sustainable and locally sourced materials with low environmental impact.

   - **AI Integration:** AI can analyze material properties, availability, and environmental impact, providing insights to architects and builders for making informed decisions on sustainable material selection.

7. **Waste Reduction:**

- **Green Practice:** Implement construction waste management practices to minimize waste.

- **AI Integration:** AI can optimize construction processes to reduce waste by analyzing historical data, identifying inefficiencies, and suggesting improvements in resource utilization.

8. **Energy Monitoring and Management:**

- **Green Practice:** Implement energy monitoring systems for real-time energy consumption tracking.

- **AI Integration:** AI can analyze energy usage data to identify opportunities for optimization, predict maintenance needs, and recommend strategies for further energy efficiency improvements.

9. **Renewable Energy Integration:**

- **Green Practice:** Incorporate renewable energy sources like solar panels and wind turbines.

- **AI Integration:** AI can predict renewable energy production patterns, optimize the integration of renewables into the building's energy grid, and ensure efficient storage and distribution.

10. **Occupant Comfort and Well-being:**

- **Green Practice:** Design for occupant comfort and well-being with proper ventilation, natural light, and indoor air quality.

- **AI Integration:** AI can monitor and analyze indoor environmental conditions, adjusting HVAC and lighting systems to create optimal conditions for occupants while maximizing energy efficiency.

11. **Commissioning and Performance Verification:**

- **Green Practice:** Conduct commissioning processes to verify that building systems operate as intended.

- **AI Integration:** AI can automate commissioning tasks, analyze performance data, and continuously optimize building systems for long-term energy efficiency.

By combining green building practices with AI technologies, construction projects can achieve higher levels of energy efficiency, sustainability, and overall environmental performance. The integration of AI allows for dynamic and adaptive systems that respond to changing conditions, contributing to the long-term resilience and eco-friendliness of buildings.

## 11.3 Renewable Energy Integration

Integrating renewable energy sources with AI in civil engineering and construction is a powerful approach to enhance energy efficiency and sustainability.

Here's how AI can be applied in the context of renewable energy integration:

1. **Resource Assessment and Prediction:**

   - **Renewable Energy Source:** Solar, wind, geothermal, etc.

   - **AI Integration:** AI can analyze historical weather data, satellite imagery, and other relevant information to predict renewable energy resource availability. This helps in optimizing the planning and design of renewable energy systems.

2. **Optimized System Design:**

- **Renewable Energy Source:** Solar panels, wind turbines, etc.

- **AI Integration:** Use AI algorithms to optimize the placement and design of renewable energy systems based on real-time and predicted environmental conditions. This ensures maximum energy capture and efficiency.

3. **Energy Storage Optimization:**

- **Renewable Energy Source:** Batteries, pumped storage, etc.

- **AI Integration:** AI can predict energy demand patterns, weather conditions, and system performance to optimize the charging and discharging cycles of energy storage systems. This maximizes the use of stored renewable energy during periods of high demand or low renewable resource availability.

4. **Grid Integration and Demand Response:**

- **Renewable Energy Source:** Distributed energy resources (DERs), smart grids.

- **AI Integration:** AI can analyze data from various sources, including energy demand, supply, and grid conditions, to optimize the integration of renewable energy into the grid. AI-driven demand response systems can dynamically adjust energy consumption patterns to match renewable energy availability.

5. **Predictive Maintenance for Renewable Assets:**

- **Renewable Energy Source:** Wind turbines, solar panels, etc.

- **AI Integration:** Implement predictive maintenance using AI to analyze data from sensors and performance metrics. This helps in detecting potential issues before they become critical, maximizing the operational lifespan of renewable energy assets.

6. **Dynamic Energy Management:**

- **Renewable Energy Source:** Solar, wind, etc.

- **AI Integration:** AI can dynamically manage the energy produced by renewable sources in real-time. By considering factors like weather conditions, energy demand, and storage capacity, AI algorithms can optimize the distribution and utilization of renewable energy within a building or across a network of buildings.

7. **Energy Forecasting:**

- **Renewable Energy Source:** Solar, wind, etc.

- **AI Integration:** AI models can forecast renewable energy production with high accuracy. These forecasts enable better planning for energy consumption and storage, ensuring that buildings can rely on renewable sources as much as possible.

8. **Integration with Building Energy Management Systems (BEMS):**

- **Renewable Energy Source:** Solar panels, geothermal systems, etc.

- **AI Integration:** AI can work in conjunction with BEMS to optimize the utilization of renewable energy within buildings. This includes adjusting heating, cooling, and lighting systems based on real-time renewable energy availability.

9. **Financial Optimization:**

- **Renewable Energy Source:** Solar, wind, etc.

- **AI Integration:** AI algorithms can analyze financial data, energy market conditions, and government incentives to optimize the financial returns of renewable energy investments. This includes determining the best times to sell excess energy back to the grid or store it for later use.

10. **Environmental Impact Assessment:**

- **Renewable Energy Source:** Solar, wind, etc.

- **AI Integration:** AI can assess the environmental impact of renewable energy projects by considering factors such as carbon footprint, land use, and ecosystem impact. This information helps in making informed decisions about the sustainability of renewable energy initiatives.

The integration of AI with renewable energy in civil engineering and construction enhances the overall efficiency, reliability, and sustainability of energy systems. By leveraging AI technologies, stakeholders can make data-driven decisions to optimize renewable energy utilization, reduce environmental impact, and contribute to a more sustainable built environment.

## 11.4 Life Cycle Assessment with AI

Life Cycle Assessment (LCA) is a crucial tool for evaluating the environmental impact of products and processes throughout their entire life cycle.

When combined with AI in the context of energy efficiency and sustainability in civil engineering and construction, LCA becomes a more powerful and efficient process. Here's how AI can be integrated into Life Cycle Assessment:

1. **Data Collection and Analysis:**

    - **LCA Stage:** Inventory analysis.

- **AI Integration:** AI can automate the collection and analysis of vast amounts of data related to raw material extraction, manufacturing processes, transportation, construction, operation, and end-of-life. Machine learning algorithms can identify patterns and trends in the data to improve accuracy and efficiency.

2. **Material Selection:**

- **LCA Stage:** Impact assessment.

- **AI Integration:** AI can assist in the selection of materials by analyzing LCA data along with other factors such as cost, availability, and performance. Machine learning models can identify optimal material choices that align with sustainability goals.

3. **Design Optimization:**

- **LCA Stage:** System optimization.

- **AI Integration:** AI can optimize building and infrastructure design based on LCA data. Machine learning algorithms can analyze different design scenarios to identify the most environmentally friendly options, considering factors such as energy efficiency, material use, and end-of-life considerations.

4. **Predictive Maintenance for Longevity:**

- **LCA Stage:** Service life and maintenance.

- **AI Integration:** AI can predict maintenance needs for different components of a structure, ensuring that it operates efficiently throughout its intended service life. This predictive maintenance approach minimizes the need for premature replacements, reducing the environmental impact associated with replacement materials and construction.

5. **Energy Performance Optimization:**

- **LCA Stage:** Use and operation.

- **AI Integration:** AI can optimize the energy performance of buildings by analyzing real-time data on energy consumption, user behavior, and environmental conditions. This information can inform adjustments to building systems, ensuring energy efficiency throughout the operational phase.

6. **Waste Reduction Strategies:**

- **LCA Stage:** End-of-life and disposal.

- **AI Integration:** AI can identify strategies to reduce waste during construction and demolition based on LCA data. Machine learning algorithms can analyze historical data to optimize construction processes and minimize waste generation.

7. **Supply Chain Optimization:**

- **LCA Stage:** Supply chain analysis.

- **AI Integration:** AI can optimize supply chains by analyzing LCA data related to the environmental impact of raw materials, transportation, and manufacturing processes. Machine learning models can identify sustainable suppliers and logistics strategies.

8. **Regulatory Compliance:**

- **LCA Stage:** Interpretation.

- **AI Integration:** AI can assist in interpreting complex regulatory requirements and ensuring compliance. This includes analyzing LCA data to demonstrate conformity with environmental standards and certifications.

9. **Continuous Improvement:**

- **LCA Stage:** Improvement assessment.

- **AI Integration:** AI can support continuous improvement efforts by analyzing ongoing LCA data and identifying areas for optimization. This iterative approach ensures that sustainability goals are met and exceeded over time.

10. **Scenario Analysis:**

- **LCA Stage:** Interpretation and improvement assessment.

- **AI Integration:** AI can perform scenario analysis by simulating different conditions and variations to predict the potential environmental impact of design and construction decisions. This helps in making informed choices that align with sustainability objectives.

The integration of AI into Life Cycle Assessment processes in civil engineering and construction facilitates more accurate, efficient, and data-driven decision-making. It enables stakeholders to identify and prioritize sustainable practices throughout the life cycle of infrastructure projects, ultimately contributing to a more environmentally responsible built environment.

# 11.5 AI for Sustainable Construction

Artificial Intelligence (AI) offers numerous opportunities to enhance sustainability in the construction industry.

By leveraging AI technologies, stakeholders can optimize processes, improve resource efficiency, reduce environmental impact, and contribute to the development of more sustainable construction practices. Here are several ways AI can be applied in sustainable construction:

1. **Energy-Efficient Building Design:**

   - AI Application: AI algorithms can analyze historical data, weather patterns, and building performance metrics to optimize architectural and HVAC system designs for energy efficiency.

2. **Material Selection and Optimization:**

- AI Application: Machine learning can analyze data on material properties, environmental impact, and cost to recommend sustainable material choices, considering factors such as embodied carbon and life cycle assessments.

3. **Construction Waste Reduction:**

- AI Application: AI can optimize construction processes to minimize waste by analyzing historical data, identifying inefficiencies, and suggesting improvements in resource utilization.

4. **Predictive Maintenance:**

- AI Application: AI-powered predictive maintenance systems can monitor equipment and infrastructure, predicting potential failures before they occur. This reduces downtime, minimizes resource consumption for replacements, and extends the lifespan of assets.

5. **Renewable Energy Integration:**

- AI Application: AI can optimize the integration of renewable energy sources into construction projects. This includes predicting energy production, managing energy storage, and dynamically adjusting energy distribution based on real-time conditions.

6. **Smart Construction Site Management:**

- AI Application: AI can enhance construction site management by analyzing data from sensors, cameras, and other sources to optimize logistics, monitor worker safety, and reduce energy consumption on-site.

7. **Optimized Project Scheduling:**

- AI Application: AI algorithms can analyze project schedules, resource availability, and external factors to optimize construction timelines. This can lead to more efficient use of resources and reduced energy consumption.

8. **Green Building Certification Support:**

- AI Application: AI can assist in the documentation and analysis required for green building certifications. Automated systems can streamline the process of gathering and presenting data to meet sustainability standards.

9. **Dynamic Building Systems Optimization:**

   - AI Application: AI can continuously analyze data from building systems (HVAC, lighting, etc.) to optimize their performance in real-time, adapting to changing conditions and occupancy patterns.

10. **Water Management:**

    - AI Application: AI can optimize water usage in construction processes and building operation by analyzing data on water consumption, predicting demand, and identifying opportunities for conservation.

11. **Carbon Footprint Monitoring:**

    - AI Application: AI can track and analyze the carbon footprint of construction projects, providing real-time insights into emissions. This information can guide decision-making to reduce environmental impact.

12. **Environmental Impact Assessment:**

    - AI Application: AI can assist in assessing the overall environmental impact of construction projects by analyzing data related to air and water quality, noise levels, and biodiversity. This information helps in implementing measures to mitigate negative impacts.

13. **Supply Chain Optimization:**

    - AI Application: AI can optimize the construction supply chain by analyzing data on suppliers, transportation, and materials, ensuring that sustainable and ethical practices are prioritized.

14. **Occupant Comfort and Well-being:**

    - AI Application: AI can monitor indoor environmental conditions, such as air quality and temperature, to ensure occupant comfort while minimizing energy consumption.

### 15. Community Engagement and Social Sustainability:

- AI Application: AI can analyze social data and community feedback to inform construction projects about the social impact and preferences of local communities, contributing to socially sustainable practices.

The integration of AI in sustainable construction practices has the potential to revolutionize the industry, making processes more efficient, cost-effective, and environmentally friendly. It enables stakeholders to make data-driven decisions that align with sustainability goals and contribute to the creation of a more resilient and eco-friendly built environment.

# 11.6 Real -world Examples / Use Cases

1. **AI in Energy Efficiency and Sustainability:**

   - **AI-powered Energy Management (Company: BrainBox AI):**

     BrainBox AI applies artificial intelligence to optimize energy efficiency in buildings. Their platform uses AI algorithms to predict and control HVAC (Heating, Ventilation, and Air Conditioning) systems, leading to significant energy savings.

   - **Energy Consumption Analytics (Company: Verdigris):**

     Verdigris utilizes AI for energy consumption analytics in commercial buildings. The AI platform analyzes data from electrical systems to identify energy inefficiencies and provides actionable insights to improve energy performance.

2. **Energy Modeling:**

   - **Energy Simulation Software (Company: Autodesk Insight):**

     Autodesk Insight offers energy simulation software for civil engineering and construction projects. The software utilizes building information modeling (BIM) data to perform energy analysis, allowing designers to optimize energy performance during the design phase.

   - **Building Performance Modeling (Company: IES Virtual Environment):**

     IES Virtual Environment provides software for building performance modeling. The platform enables architects and engineers to simulate and analyze the energy performance of buildings, informing design decisions for enhanced sustainability.

3. **Green Building Practices:**

   - **LEED Certification Support (Company: Arc Skoru):**

Arc Skoru supports green building practices by offering tools for LEED (Leadership in Energy and Environmental Design) certification. Their platform uses data analytics to assess and improve the sustainability performance of buildings.

- **Sustainable Design Software (Company: SketchUp):**

    SketchUp provides sustainable design software for architects and designers. The software allows users to create environmentally conscious designs by integrating energy-efficient features and sustainable materials.

4. **Renewable Energy Integration:**

- **Smart Microgrid Solutions (Company: Siemens):**

    Siemens offers smart microgrid solutions for renewable energy integration. Their systems use AI to optimize the integration of renewable energy sources, ensuring efficient and sustainable power distribution in buildings and communities.

- **Solar Energy Optimization (Company: Aurora Solar):**

    Aurora Solar utilizes AI for solar energy optimization in the design and planning of solar installations. The platform analyzes factors such as shading and sunlight exposure to optimize the placement of solar panels for maximum energy generation.

5. **Life Cycle Assessment with AI:**

- **AI-driven Life Cycle Assessment (Company: One Click LCA):**

    One Click LCA integrates AI for life cycle assessment in civil engineering and construction projects. The platform assesses the environmental impact of buildings and infrastructure, helping stakeholders make informed decisions for sustainability.

- **Environmental Impact Analysis (Company: SimaPro):**

SimaPro provides software for environmental impact analysis and life cycle assessment. The platform employs AI to analyze data on material use, energy consumption, and emissions, supporting sustainable decision-making in construction projects.

6. **AI for Sustainable Construction:**

   - **Sustainable Construction Analytics (Company: Built Green Canada):**

     Built Green Canada utilizes AI for sustainable construction analytics. The organization employs AI algorithms to analyze construction practices and materials, promoting sustainable and environmentally friendly building practices.

   - **AI-driven Green Certification (Company: Green Badger):**

     Green Badger offers AI-driven solutions for green certification in construction. The platform assists in achieving and managing green building certifications by automating compliance tracking and reporting.

# 11.7 Chapter Summary: Key Points

1. AI optimizes energy consumption in buildings by analyzing sensor data, adjusting lighting, heating, and cooling in real-time for efficiency and occupant comfort.

2. In renewable energy, AI predicts production patterns, optimizes operation, and integrates sources into the power grid, reducing reliance on non-renewable energy.

3. AI-driven predictive maintenance enhances energy infrastructure reliability and lifespan by analyzing sensor data to predict faults.

4. Smart grids, powered by AI, optimize electricity distribution based on real-time demand, reducing wastage and enhancing overall efficiency.

5. AI in energy analytics identifies opportunities for efficiency improvements by analyzing historical energy data.

6. AI optimizes energy storage systems by predicting storage needs, balancing supply and demand, and reducing grid stress.

7. In transportation, AI optimizes traffic flow, reduces congestion, and minimizes fuel consumption through smart traffic management systems.

8. Intelligent energy management systems in industries use AI to analyze processes, optimize energy use, and reduce waste.

9. AI in environmental monitoring analyzes sensor and satellite data for ecosystems, tracking changes and supporting sustainable resource management.

10. Energy modeling in civil engineering integrates AI for building performance optimization, predictive maintenance, and smart grids.

11. Green building practices benefit from AI in site selection, energy-efficient design, HVAC systems optimization, and material selection.

12. AI enhances renewable energy integration by predicting resource availability, optimizing system design, and managing energy storage and grid integration.

13. AI in Life Cycle Assessment automates data collection, aids material selection, optimizes design, and supports continuous improvement for sustainable construction.

14. AI applications in sustainable construction include energy-efficient design, material selection, construction waste reduction, and renewable energy integration.

15. AI contributes to dynamic building systems optimization, project scheduling, green building certification, and supply chain optimization in sustainable construction practices.

# 11.8 Concept Check: Q & A Sessions

1. How does AI contribute to optimizing energy consumption in buildings, and what role does it play in smart building systems?

2. What are the key applications of AI in advancing energy efficiency and sustainability in civil engineering and construction?

3. How does AI-driven predictive maintenance contribute to optimizing the performance of energy infrastructure?

4. In what ways do smart grids benefit from AI technologies, and how do they contribute to a more flexible and sustainable energy ecosystem?

5. How does AI in energy analytics identify opportunities for efficiency improvements, and what type of data does machine learning models analyze?

6. What role does AI play in optimizing energy storage systems, and how do machine learning models contribute to balancing supply and demand?

7. How does AI contribute to promoting sustainable practices in the transportation sector, specifically in optimizing traffic flow and minimizing fuel consumption?

8. How does AI support the development of intelligent energy management systems for industries, and what aspects do these systems analyze to optimize energy use?

9. What role does AI play in environmental monitoring and conservation efforts, and what types of data does it analyze?

10. How does AI contribute to collaborative design and decision-making in civil engineering, and what datasets does it analyze to provide insights for energy-efficient and sustainable design decisions?

# 12. Human-Machine Collaboration in Construction

Human-machine collaboration in the construction industry represents a transformative approach that leverages the strengths of both humans and machines to enhance productivity, safety, and overall project outcomes.

Construction sites are dynamic environments, and the integration of machines, equipped with advanced technologies, augments the capabilities of human workers.

Machines, such as construction robots and autonomous vehicles, are employed for tasks that require precision, strength, and efficiency. These machines can execute repetitive and labor-intensive activities, such as heavy lifting, excavation, and material transportation, reducing the physical strain on human workers and mitigating the risk of workplace injuries.

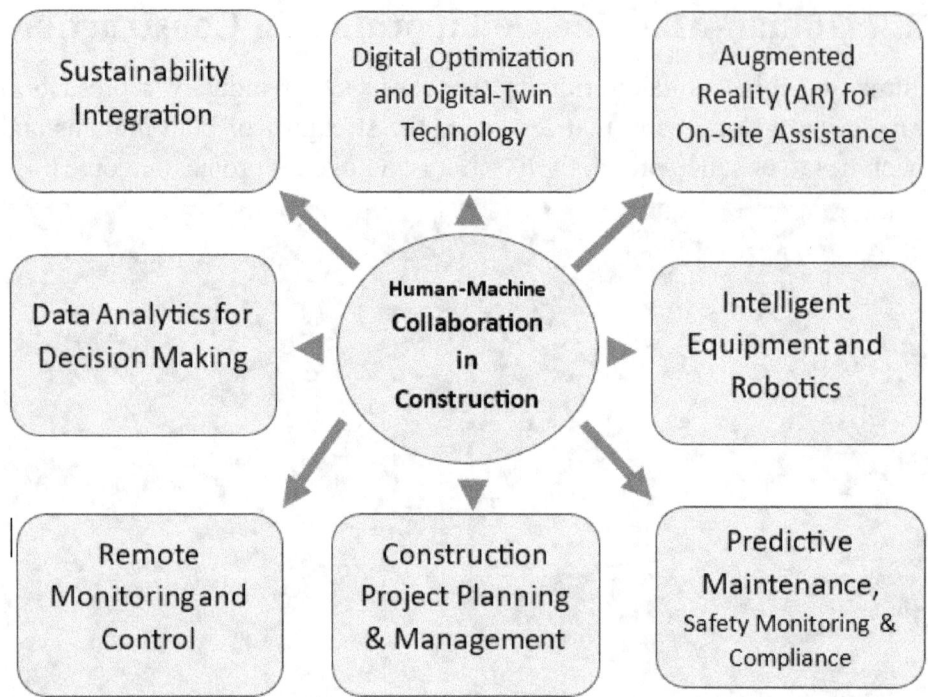

Human-machine collaboration in construction is revolutionizing the industry by integrating advanced technologies to optimize various aspects of the construction process. This collaboration involves leveraging generative design tools within collaborative platforms to empower architects and engineers to create optimized designs that balance aesthetic, functional, and construction considerations. Digital twin technology enables real-time monitoring and simulation, allowing stakeholders to visualize and analyze construction projects, identify potential issues, and make informed decisions. Augmented Reality (AR) applications provide on-site workers with real-time information, enhancing collaboration, reducing errors, and improving task efficiency. Collaborating with intelligent, autonomous machinery, such as excavators and bricklayers, improves construction speed and accuracy while ensuring human operators work alongside machines for safety. Predictive maintenance, AI-driven planning, remote monitoring, and data analytics contribute to more efficient project execution, cost control, and timely delivery. Safety is enhanced through collaborative efforts involving wearable technology for on-site workers, and sustainability is promoted by collaborating with AI systems to analyze data related to material selection, energy usage, and waste management. Overall, human-machine collaboration in construction maximizes efficiency, safety, and sustainability, contributing to the success of construction projects.

Drones equipped with cameras and sensors contribute to human-machine collaboration by providing aerial surveys and real-time data collection. This technology aids in monitoring construction sites, assessing progress, and identifying potential issues, enabling more informed decision-making by project managers and engineers.

Building Information Modeling (BIM) is a central component of human-machine collaboration in construction. BIM technology allows for the creation of digital representations of construction projects, facilitating collaborative planning and decision-making among architects, engineers, and construction teams. This shared digital environment enhances communication and coordination throughout the project lifecycle.

Wearable technologies, such as smart helmets and exoskeletons, enhance the capabilities of human workers on construction sites. These devices provide real-time information, augment physical strength, and improve safety by monitoring factors like temperature and worker fatigue.

Artificial Intelligence (AI) algorithms contribute to human-machine collaboration by analyzing vast amounts of data from construction sites. AI can predict potential issues, optimize project schedules, and assist in resource allocation, helping project managers make data-driven decisions for efficient project execution.

Human workers contribute their cognitive skills, creativity, and problem-solving abilities to the collaborative process. While machines excel in repetitive and physically demanding tasks, humans bring adaptability, intuition, and contextual understanding to the construction site, addressing challenges that may arise in complex and unpredictable situations.

Telematics and Internet of Things (IoT) devices are integrated into construction equipment, providing real-time data on machine performance and location. This information aids in fleet management, preventive maintenance, and optimizing the utilization of construction machinery, leading to increased operational efficiency.

Training and skill development are crucial aspects of human-machine collaboration. Workers need to be trained to operate and collaborate with advanced technologies. This requires ongoing education programs to ensure that the workforce is equipped with the skills needed to navigate and leverage the capabilities of evolving construction technologies.

Ethical considerations, including data privacy, job displacement, and worker safety, are integral to human-machine collaboration. Striking a balance between automation and human involvement, ensuring fair treatment of workers, and addressing concerns related to job security are essential for the responsible integration of machines into the construction industry.

In conclusion, human-machine collaboration in construction fosters a symbiotic relationship where the strengths of humans and machines complement each other. This collaborative approach is driving innovation, improving efficiency, and creating safer and more productive construction environments, ultimately shaping the future of the construction industry.

## 12.1 Augmented Intelligence

Augmented Intelligence (AI) in the context of human-machine collaboration in construction refers to the use of technology to enhance and amplify human capabilities, rather than replacing them.

Here are key aspects of how augmented intelligence is applied in the construction industry to facilitate collaboration between humans and machines:

1. **Design and Planning:**

   - **Human Expertise:** Architects and engineers bring design creativity and domain expertise.

- **AI Augmentation:** AI tools assist in generating design alternatives, conducting simulations, and analyzing data for informed decision-making during the planning phase. This collaborative approach enhances efficiency and design optimization.

2. **Building Information Modeling (BIM):**

   - **Human Expertise:** Construction professionals interpret and implement BIM models.

   - **AI Augmentation:** AI aids in automating repetitive tasks within BIM, such as clash detection, quantity take-offs, and model optimization. This allows humans to focus on higher-level decision-making and creative aspects of the project.

3. **Project Management:**

   - **Human Expertise:** Project managers provide leadership, communication, and strategic planning.

   - **AI Augmentation:** AI tools assist in project scheduling, resource allocation, risk management, and performance monitoring. This collaboration enhances project efficiency and helps in identifying potential issues before they escalate.

4. **Construction Automation:**

   - **Human Expertise:** Skilled workers perform complex and specialized construction tasks.

   - **AI Augmentation:** Automation technologies, guided by AI, assist in routine and repetitive tasks, such as bricklaying, concrete pouring, or welding. Human workers collaborate with automated systems to improve productivity and safety.

5. **Safety Monitoring:**

   - **Human Expertise:** Safety officers and workers are responsible for on-site safety.

   - **AI Augmentation:** AI-powered sensors and cameras monitor construction sites for potential safety hazards. Augmented intelligence provides real-time alerts and assists in proactively addressing safety concerns.

6. **Quality Control:**

- **Human Expertise:** Quality inspectors evaluate construction work for compliance.

- **AI Augmentation:** AI-based tools analyze construction elements for quality control, detecting deviations from specifications. Human inspectors can then focus on critical and nuanced assessments.

7. **Supply Chain Management:**

- **Human Expertise:** Procurement professionals manage the supply chain and vendor relationships.

- **AI Augmentation:** AI helps optimize supply chain operations by predicting material requirements, assessing supplier performance, and identifying cost-saving opportunities. Humans make strategic decisions based on AI-driven insights.

8. **Predictive Maintenance:**

- **Human Expertise:** Maintenance teams perform routine inspections and repairs.

- **AI Augmentation:** AI predicts equipment failures, recommends maintenance schedules, and helps optimize asset performance. Human workers collaborate with AI to ensure equipment reliability.

9. **Collaborative Robotics (Cobots):**

- **Human Expertise:** Skilled workers collaborate with robots on the construction site.

- **AI Augmentation:** AI-powered collaborative robots (cobots) work alongside humans, assisting in tasks that require precision, strength, or repetitive actions. This enhances productivity and worker safety.

10. **Data Analytics for Decision-Making:**

- **Human Expertise:** Project managers and executives make strategic decisions.

- **AI Augmentation:** AI analyzes vast amounts of data to provide insights for decision-making. Humans interpret the results, consider broader implications, and make informed choices for project success.

**11. Communication and Collaboration Platforms:**

- **Human Expertise:** Project teams communicate and collaborate.

- AI Augmentation: AI-driven communication platforms facilitate seamless collaboration by automating routine tasks, managing workflows, and providing intelligent insights. This helps streamline communication and coordination.

12. **Learning and Training:**

- **Human Expertise:** Skilled workers provide on-the-job training.

- **AI Augmentation:** AI-driven training platforms use simulations and virtual reality to enhance learning experiences for construction workers. This collaborative approach accelerates skills development and improves overall workforce competency.

In summary, augmented intelligence in construction focuses on empowering human professionals with AI tools and technologies to enhance productivity, safety, and decision-making. This collaborative approach leverages the strengths of both humans and machines, resulting in more efficient and effective construction processes.

## 12.2 Cognitive Computing

Cognitive computing in the context of human-machine collaboration in construction involves systems that can understand, learn, and interact with users in a natural and intelligent way.

These technologies aim to enhance collaboration by augmenting human cognitive abilities and supporting decision-making processes in the construction industry. Here are key aspects of how cognitive computing contributes to human-machine collaboration in construction:

1.  **Natural Language Processing (NLP):**

    *   **Human Interaction:** Project managers, engineers, and workers communicate in natural language.

- **Cognitive Computing Integration:** NLP in cognitive computing enables systems to understand and respond to human language. This facilitates seamless communication between construction professionals and AI-driven systems, improving information exchange and collaboration.

2. **Context-Aware Systems:**

   - **Human Interaction:** Construction professionals work in dynamic and evolving environments.

   - **Cognitive Computing Integration:** Cognitive systems can analyze contextual data from construction sites, such as weather conditions, project schedules, and safety parameters. This information enhances decision-making by considering real-time factors that impact construction processes.

3. **Machine Learning for Predictive Analytics:**

   - **Human Interaction:** Project managers make decisions based on historical data and experience.

   - **Cognitive Computing Integration:** Machine learning algorithms in cognitive computing analyze historical data to predict project outcomes, potential delays, and resource requirements. This supports project managers in making informed decisions and planning effectively.

4. **Computer Vision for Visual Analysis:**

   - **Human Interaction:** Engineers and inspectors visually assess construction sites.

   - **Cognitive Computing Integration:** Computer vision technologies in cognitive computing enable systems to analyze visual data from construction sites. This includes monitoring progress, identifying safety hazards, and conducting quality control assessments, enhancing human visual inspections.

5. **Knowledge Graphs for Data Integration:**

   - **Human Interaction:** Construction professionals access information from various sources.

- **Cognitive Computing Integration:** Knowledge graphs in cognitive computing organize and connect diverse data sources, providing a holistic view of project information. This supports construction professionals in accessing relevant data for decision-making and collaboration.

6. **Expert Systems for Decision Support:**

   - **Human Interaction:** Construction managers make decisions based on expertise.

   - **Cognitive Computing Integration:** Expert systems in cognitive computing leverage accumulated knowledge and best practices to provide decision support. Construction managers can receive recommendations and insights to enhance their decision-making process.

7. **Autonomous and Semi-Autonomous Machinery:**

   - **Human Interaction:** Equipment operators control construction machinery.

   - **Cognitive Computing Integration:** Autonomous and semi-autonomous machinery equipped with cognitive computing capabilities can analyze real-time data to optimize operations, navigate construction sites, and ensure safety. Operators collaborate with these intelligent systems for more efficient and safe construction processes.

8. **Virtual and Augmented Reality (VR/AR):**

   - **Human Interaction:** Designers and workers visualize projects using VR/AR.

   - **Cognitive Computing Integration:** Cognitive computing enhances VR/AR experiences by providing intelligent insights and real-time information overlays. This supports collaborative design reviews, on-site navigation, and decision-making based on augmented information.

9. **Emotion Recognition:**

   - **Human Interaction:** Team members express emotions during project collaboration.

- **Cognitive Computing Integration:** Emotion recognition technologies in cognitive computing can analyze facial expressions and other cues to understand the emotional state of construction professionals. This insight can be used to improve team dynamics and address issues promptly.

## 10. Dynamic Scheduling and Resource Allocation:

- **Human Interaction:** Project managers allocate resources and plan schedules.

- **Cognitive Computing Integration:** Cognitive systems use AI algorithms to dynamically adjust project schedules and resource allocations based on real-time data. This collaborative approach ensures optimal resource utilization and minimizes delays.

## 11. Continuous Learning and Adaptation:

- **Human Interaction:** Construction professionals adapt to evolving project requirements.

- **Cognitive Computing Integration:** Cognitive systems continuously learn from new data and user interactions. This adaptive capability ensures that AI-driven solutions evolve alongside changing project conditions and requirements.

In the construction industry, cognitive computing supports human-machine collaboration by providing intelligent insights, automating routine tasks, and enhancing decision-making processes. The combination of human expertise and cognitive computing capabilities contributes to more efficient, informed, and collaborative construction projects.

## 12.3 Collaborative Robots (Cobots)

Collaborative Robots, or Cobots, play a significant role in fostering human-machine collaboration in the construction industry.

Unlike traditional industrial robots that typically work in isolation, Cobots are designed to work alongside human workers, enhancing efficiency, safety, and flexibility in construction processes. Here's how Cobots contribute to human-machine collaboration in construction:

1. **Construction Tasks Assistance:**

   - **Human Interaction:** Construction workers perform various tasks on-site.

- **Cobot Integration:** Cobots can assist human workers in tasks such as heavy lifting, material transportation, and repetitive activities. This collaborative approach improves overall productivity and reduces the physical strain on human workers.

2. **Precision and Accuracy:**

   - **Human Interaction:** Skilled workers perform tasks that require precision.

   - **Cobot Integration:** Cobots equipped with advanced sensors and vision systems can work with high precision. They can assist in tasks such as bricklaying, welding, or concrete pouring, ensuring accuracy and quality in construction processes.

3. **Safety Enhancement:**

   - **Human Interaction:** Construction sites pose various safety challenges.

   - **Cobot Integration:** Cobots are designed with built-in safety features, including sensors that detect the presence of humans and stop or slow down if they come too close. This minimizes the risk of accidents and enhances overall safety on construction sites.

4. **Collaborative Design and Planning:**

   - **Human Interaction:** Architects and engineers collaborate on design and planning.

   - **Cobot Integration:** Cobots can assist in collaborative design processes by simulating and visualizing construction scenarios. They contribute to virtual planning, allowing human professionals to optimize designs and address potential challenges.

5. **Site Inspection and Monitoring:**

   - **Human Interaction:** Inspectors and supervisors monitor construction progress.

   - **Cobot Integration:** Cobots equipped with cameras and sensors can conduct site inspections, monitor construction progress, and collect data on project status. This collaborative monitoring assists human supervisors in making informed decisions.

6.  **Flexible Task Execution:**

    - **Human Interaction:** Construction tasks often require adaptability.

    - **Cobot Integration:** Cobots are programmable and can be easily reconfigured to perform different tasks. Human workers can collaborate with Cobots to quickly adapt to changing project requirements or address unexpected challenges on-site.

7.  **Tight Spaces and Hazardous Environments:**

    - **Human Interaction:** Certain construction tasks occur in confined or hazardous spaces.

    - **Cobot Integration:** Cobots can access tight spaces and work in hazardous environments, reducing the need for human workers to expose themselves to potential risks. This collaborative approach enhances safety and efficiency in challenging conditions.

8.  **Training and Skill Development:**

    - **Human Interaction:** Workers need training for specific construction tasks.

    - **Cobot Integration:** Cobots can be used for training purposes, allowing workers to interact with them in a controlled environment to gain experience and skills. This collaborative training approach accelerates the learning curve for new tasks.

9.  **Demolition and Deconstruction:**

    - **Human Interaction:** Demolition and deconstruction involve physical labor and safety risks.

    - **Cobot Integration:** Cobots equipped with tools for demolition tasks can work alongside human workers to safely and efficiently dismantle structures. This collaborative approach minimizes risks associated with manual demolition.

10. **Logistics and Material Handling:**

- **Human Interaction:** Material transportation and logistics require physical effort.

- **Cobot Integration:** Cobots can be used for material handling tasks, transporting materials across construction sites. Human workers collaborate with Cobots to optimize material flow and reduce manual labor.

## 11. Data Collection and Analysis:

- **Human Interaction:** Workers collect and analyze data for decision-making.

- **Cobot Integration:** Cobots equipped with sensors and cameras can collect real-time data on construction processes. Human workers collaborate with Cobots to analyze this data for decision-making, improving project efficiency.

In summary, collaborative robots play a crucial role in human-machine collaboration in construction by assisting with various tasks, enhancing precision, improving safety, and contributing to overall project efficiency. The synergy between human workers and Cobots represents a key aspect of the evolving technological landscape in the construction industry.

## 12.4 Skills and Training for AI Collaboration

In the context of human-machine collaboration in construction, individuals need a combination of technical skills and adaptability to work effectively with AI technologies.

Here are key skills and training areas for professionals engaging in AI collaboration in the construction industry:

1. **Digital Literacy:**

   - **Skills:** Understanding basic computer operations, software usage, and familiarity with digital tools.

- **Training:** Basic computer literacy courses, online tutorials, and workshops on using digital tools commonly employed in construction projects.

2. **Data Literacy:**

- **Skills:** Ability to understand, interpret, and work with data.

- **Training:** Data literacy courses that cover data analysis, interpretation, and the use of data visualization tools. Understanding how data drives decision-making in construction projects.

3. **Programming Basics:**

- **Skills:** Basic programming knowledge to understand and interact with AI systems.

- **Training:** Introductory courses in programming languages relevant to AI applications, such as Python or JavaScript.

4. **AI Awareness:**

- **Skills:** Knowledge of AI concepts, terminology, and applications in construction.

- **Training:** AI awareness programs, workshops, or online courses that introduce the fundamentals of AI and its potential applications in the construction industry.

5. **Collaboration Skills:**

- **Skills:** Ability to work effectively in interdisciplinary teams that include both humans and AI systems.

- **Training:** Team-building exercises, workshops on effective communication, and collaborative problem-solving scenarios.

6. **Problem-Solving:**

- **Skills:** Critical thinking and creative problem-solving skills to address challenges in construction projects.

- **Training:** Problem-solving workshops, case studies, and exercises that encourage individuals to think critically and find innovative solutions.

7. **Adaptability:**

- **Skills:** Ability to adapt to changing technologies and evolving project requirements.

- **Training:** Continuous learning programs, workshops on adapting to new technologies, and exposure to emerging trends in construction tech.

8. **Technical Training on AI Tools:**

- **Skills:** Proficiency in using AI tools and platforms specific to construction applications.

- **Training:** Hands-on training programs focused on using AI tools relevant to construction tasks, such as BIM software, AI-driven scheduling systems, or predictive maintenance platforms.

9. **Robotics and Automation Skills:**

- **Skills:** Familiarity with robotics and automated systems used in construction.

- **Training:** Robotics workshops, training programs on operating and collaborating with robotic systems, and hands-on experience with automation tools.

10. **Ethics and Responsible AI:**

- **Skills:** Understanding ethical considerations related to AI use in construction.

- **Training:** Courses on ethics in AI, workshops on responsible AI practices, and discussions on the societal impacts of AI in construction.

11. **Security Awareness:**

- **Skills:** Awareness of cybersecurity and data privacy concerns related to AI collaboration.

- **Training:** Cybersecurity workshops, courses on data protection, and awareness programs on securing AI systems in construction.

12. **Continuous Learning and Professional Development:**

- **Skills:** A commitment to ongoing learning and staying updated on AI advancements.

- **Training:** Encouraging and supporting employees in pursuing certifications, attending conferences, and participating in industry events to stay abreast of the latest AI developments.

### 13. Soft Skills:

- **Skills:** Interpersonal skills, emotional intelligence, and effective communication.

- **Training:** Soft skills workshops, communication training, and team-building exercises to foster positive collaboration among team members.

### 14. Health and Safety Training:

- **Skills:** Understanding health and safety considerations when working with AI systems on construction sites.

- **Training:** Health and safety courses, workshops on AI-related safety practices, and protocols for working with AI-enabled machinery.

By cultivating these skills and providing relevant training, individuals can effectively collaborate with AI technologies in the construction industry, fostering a harmonious integration of human and machine capabilities. This approach contributes to the overall success of construction projects and ensures a skilled workforce ready to embrace the benefits of AI collaboration.

## 12.5 Future Workforce Dynamics

The future workforce dynamics for human-machine collaboration in construction will be shaped by technological advancements, evolving industry demands, and the need for a skilled and adaptable workforce.

Here are several key aspects that will influence the future of human-machine collaboration in the construction industry:

1. **Increased Integration of AI and Robotics:**

   - **Expectation:** Greater integration of AI and robotic systems into construction processes.

   - **Impact:** Construction professionals will need to be adept at collaborating with and overseeing AI-driven systems and robotic equipment. Training programs will focus on understanding and leveraging advanced technologies for enhanced productivity and efficiency.

2. **Skills Shift:**

   - **Expectation:** Shift in the required skill set towards digital and technical competencies.

   - **Impact:** Construction workers and professionals will need to acquire skills related to data analysis, programming, and the operation of AI-driven tools. Upskilling and reskilling initiatives will be essential to keep pace with technological advancements.

3. **Rise of Remote Collaboration:**

   - **Expectation:** Increased use of remote collaboration tools and platforms.

   - **Impact:** Construction professionals may collaborate with remote teams, leveraging virtual reality, augmented reality, and collaborative software. Skills related to virtual collaboration, digital communication, and remote project management will become crucial.

4. **Interdisciplinary Collaboration:**

   - **Expectation:** Greater collaboration between construction professionals and experts from diverse fields.

   - **Impact:** Cross-disciplinary collaboration will be essential for successful project outcomes. Professionals with the ability to communicate effectively across disciplines, including technology specialists, will be highly valued.

5. **Continued Emphasis on Safety:**

   - **Expectation:** Continued focus on enhancing safety through AI-driven technologies.

   - **Impact:** Safety training will include AI-specific considerations, and workers will collaborate with AI systems to ensure safe working conditions. The integration of AI into safety protocols will become standard practice.

6. **Flexible and Adaptive Workforce:**

- **Expectation:** Need for a flexible and adaptive workforce to navigate changing technologies.

- **Impact:** Construction professionals will need to embrace a mindset of continuous learning and adaptability. Organizations will invest in training programs to foster a culture of innovation and agility.

7. **Job Evolution, Not Elimination:**

- **Expectation:** Transformation of job roles rather than wholesale job elimination.

- **Impact:** While some routine tasks may be automated, new roles will emerge that require human skills such as critical thinking, problem-solving, creativity, and emotional intelligence. Workers will be involved in overseeing, optimizing, and collaborating with AI and robotic systems.

8. **Increased Emphasis on Sustainability:**

- **Expectation:** Growing focus on sustainable construction practices.

- **Impact:** Construction professionals will collaborate with AI systems to optimize designs for sustainability, reduce environmental impact, and implement eco-friendly construction processes. Green construction practices will become integral to industry standards.

9. **Data Privacy and Security Concerns:**

- **Expectation:** Greater emphasis on protecting data privacy and ensuring cybersecurity.

- **Impact:** Construction professionals will need to be aware of and address data privacy and security concerns associated with AI collaboration. Training programs will incorporate best practices for securing sensitive information.

10. **Government Regulations and Standards:**

- **Expectation:** Development of regulations and standards for AI applications in construction.

- **Impact:** Construction professionals will need to stay informed about evolving regulations related to AI use in construction. Compliance with standards will be essential to ensure the ethical and responsible deployment of AI technologies.

## 11. Cultural Shift:

- **Expectation:** Cultural shift towards embracing technological advancements in construction.

- **Impact:** Organizations will promote a culture of innovation and openness to change. Construction professionals will be encouraged to adopt a positive attitude toward AI and technology integration.

## 12. Global Collaboration:

- **Expectation:** Increased collaboration on international construction projects.

- **Impact:** Construction professionals will work on projects that involve teams from different countries. Cross-cultural communication skills and an understanding of international regulations and standards will be crucial.

In navigating these future workforce dynamics, a holistic approach to education, training, and workplace culture will be essential. The construction industry can prepare for human-machine collaboration by investing in education programs, fostering a culture of innovation, and creating opportunities for continuous learning and development. As construction becomes more digitized and AI-driven, the workforce will play a pivotal role in driving successful project outcomes and industry advancements.

## 12.6 Real-world Examples / Use Cases

1. **Human-Machine Collaboration:**

   - **BIM-based Collaboration (Company: Autodesk BIM 360):**

     Autodesk BIM 360 facilitates human-machine collaboration in construction by integrating Building Information Modeling (BIM) technology. The platform allows teams to collaborate on 3D models, improving coordination and communication throughout the construction process.

   - **Construction Management Collaboration (Company: Procore):**

     Procore provides construction management software that enhances human-machine collaboration. The platform streamlines project communication, document management, and collaboration among construction teams and stakeholders.

2. **Augmented Intelligence:**

   - **AI-enhanced Design Assistance (Company: ClearEdge3D):**

     ClearEdge3D utilizes augmented intelligence for design assistance in construction. Their platform employs AI algorithms to assist in the extraction and validation of 3D building information models, improving accuracy and efficiency in design processes.

   - **AI-driven Construction Data Analysis (Company: Smartvid.io):**

     Smartvid.io applies augmented intelligence to construction data analysis. Their platform uses machine learning algorithms to analyze visual data, identifying trends and insights that contribute to informed decision-making in construction projects.

3. **Cognitive Computing:**

   - **Cognitive Project Management (Company: IBM Watson Construction):**

IBM Watson Construction incorporates cognitive computing for project management in construction. The platform leverages AI to analyze project data, predict risks, and provide recommendations, enhancing decision-making and project outcomes.

- **AI-enhanced Document Understanding (Company: Hyperscience):**

Hyperscience employs cognitive computing for document understanding in construction. Their platform uses AI algorithms to automate the extraction and analysis of information from documents, improving document management and workflow efficiency.

4. **Collaborative Robots (Cobots):**

- **Construction Robotics (Company: Built Robotics):**

Built Robotics specializes in collaborative robots for construction. Their autonomous equipment, such as bulldozers and excavators, collaborates with human operators to enhance construction efficiency and safety.

- **Cobotic Bricklaying (Company: Construction Robotics - SAM):**

Construction Robotics developed the Semi-Automated Mason (SAM), a collaborative robot for bricklaying. SAM works alongside human masons, improving the speed and precision of bricklaying in construction.

5. **Skills and Training for AI Collaboration:**

- **AI Training for Construction Professionals (Platform: Udacity):**

Udacity offers online courses on AI and machine learning for construction professionals. These courses provide the skills and knowledge needed for professionals to effectively collaborate with AI technologies in the construction industry.

- **AI Construction Management Certification (Organization: AACE International):**

AACE International provides certifications, including Certified Forensic Claims Consultant (CFCC), that cover AI in construction management. These certifications ensure that construction professionals are equipped with the necessary skills for AI collaboration.

6. **Future Workforce Dynamics for AI:**

- **Workforce Development Platforms (Company: Procore Construction OS):**

   Procore Construction OS includes features for workforce development in the construction industry. The platform supports training and collaboration among construction teams, preparing them for the evolving dynamics of AI integration.

- **AI-driven Talent Development (Company: Skillsoft):**

   Skillsoft offers AI-driven talent development solutions for the construction workforce. Their platform provides training in emerging technologies, ensuring that workers are prepared for the future dynamics of AI in civil engineering and construction.

# 12.7 Chapter Summary: Key Points

1. Human-machine collaboration in construction optimizes productivity, safety, and project outcomes by integrating advanced technologies with the strengths of human workers.

2. Construction robots and autonomous vehicles handle precision and labor-intensive tasks, reducing physical strain on workers and minimizing workplace injuries.

3. Generative design, digital twin technology, and AR applications enhance collaboration, streamline decision-making, and improve efficiency throughout the construction process.

4. AI-driven tools contribute to predictive maintenance, data analytics, and efficient project execution, ensuring safety and promoting sustainability in construction projects.

5. Drones provide aerial surveys, real-time data, and monitoring capabilities for informed decision-making by project managers and engineers.

6. Building Information Modeling (BIM) facilitates collaborative planning and decision-making, enhancing communication and coordination among architects, engineers, and construction teams.

7. Wearable technologies improve worker capabilities, monitor safety factors, and contribute to a safer construction environment.

8. Human expertise complements AI in tasks requiring creativity, adaptability, and contextual understanding, addressing challenges in complex and unpredictable situations.

9. Telematics and IoT devices integrated into construction equipment provide real-time data, aiding in fleet management and optimizing machinery utilization.

10. Ongoing education programs are crucial for training the workforce to operate and collaborate with evolving construction technologies.

11. Ethical considerations, including data privacy, job displacement, and worker safety, are essential for responsible integration of machines into the construction industry.

12. Augmented intelligence enhances collaboration by empowering architects, engineers, and project managers with AI tools for design, planning, project management, and safety monitoring.

13. Cognitive computing supports decision-making through natural language processing, machine learning, computer vision, and autonomous machinery in construction.

14. Collaborative robots (Cobots) work alongside human workers, assisting in tasks, improving precision, ensuring safety, and contributing to overall project efficiency.

15. Skills and training for AI collaboration in construction include digital literacy, data literacy, programming basics, collaboration skills, and continuous learning for a flexible and adaptive workforce.

## 12.8 Concept Check: Q & A Sessions

1. What is human-machine collaboration in the construction industry?

2. How do drones contribute to human-machine collaboration in construction?

3. What role does Building Information Modeling (BIM) play in human-machine collaboration in construction?

4. How do wearable technologies contribute to the capabilities of human workers on construction sites?

5. What is augmented intelligence in the context of human-machine collaboration in construction?

6. How does cognitive computing contribute to decision-making processes in the construction industry?

7. What distinguishes collaborative robots (Cobots) from traditional industrial robots in construction?

8. What skills and training are crucial for professionals engaging in AI collaboration in the construction industry?

9. How will the future workforce dynamics in construction be influenced by AI and robotics?

10. What are some key expectations for the future of human-machine collaboration in construction, and how will they impact the workforce?

# 13. AI Adoption Strategies in Construction Firms

AI adoption strategies in construction firms involve a systematic approach to integrating artificial intelligence technologies into various aspects of project planning, execution, and management.

Firstly, firms need to conduct a comprehensive assessment of their current workflows, identifying areas where AI can add value. This involves understanding specific pain points, inefficiencies, and opportunities for improvement within the construction processes.

Once the areas for AI integration are identified, construction firms can formulate a clear and achievable roadmap. This involves setting specific goals, timelines, and milestones for AI adoption. Prioritizing AI applications based on their potential impact on efficiency, cost savings, and overall project success is crucial in developing a focused and effective strategy. In the construction industry, successful AI adoption relies on a strategic approach encompassing following key strategies.

Firstly, construction firms must articulate clear objectives and use cases, identifying specific challenges AI can address, and setting measurable goals. Robust data management and quality assurance are crucial, necessitating the development of a strong data infrastructure and ensuring accuracy in input data. Collaboration is fostered through cross-functional teams, involving construction experts and IT professionals, and training programs to enhance AI awareness. Starting with pilot projects allows for controlled testing and refinement before broader implementation. Seamless integration with existing systems, coupled with addressing security and ethical considerations, ensures smooth adoption. Continuous monitoring and improvement, along with adaptability to dynamic industry changes, are integral for optimizing AI applications, leading to enhanced efficiency, data-driven decision-making, and improved project outcomes.

Collaboration and communication play a central role in AI adoption. Construction firms should foster a culture that encourages collaboration between technical teams, project managers, and other stakeholders. Open communication channels help in articulating the benefits of AI adoption, addressing concerns, and ensuring a shared understanding of the technology's potential impact.

Incorporating AI into existing workflows requires investment in talent and technology. Construction firms need to identify the right skill sets for managing AI initiatives and provide training for their workforce. Additionally, selecting appropriate AI technologies and tools that align with the firm's specific needs is crucial for successful implementation.

Piloting AI initiatives on a smaller scale allows construction firms to test and validate the technology before broader adoption. Implementing pilot projects helps in identifying challenges, fine-tuning processes, and building confidence among stakeholders regarding the benefits of AI in construction workflows.

Integration with existing technologies and systems is a key consideration in AI adoption. Construction firms should ensure that AI technologies seamlessly interface with their current software, hardware, and data infrastructure. Compatibility and interoperability are essential for a smooth transition and effective utilization of AI capabilities.

Data management is a critical aspect of AI adoption in construction. Construction firms need to establish robust data governance practices, ensuring data quality, security, and accessibility. A well-organized data infrastructure forms the foundation for AI applications, allowing algorithms to operate on reliable and relevant information.

Building partnerships with AI solution providers and technology experts can accelerate the adoption process. Collaboration with industry specialists facilitates knowledge transfer, access to cutting-edge technologies, and ongoing support for overcoming challenges encountered during implementation.

Measuring and evaluating the success of AI adoption is an ongoing process. Construction firms should establish key performance indicators (KPIs) aligned with their goals and regularly assess the impact of AI on project outcomes, efficiency gains, and overall business performance.

Continuous learning and adaptation are essential for staying at the forefront of AI innovation. Construction firms should stay informed about advancements in AI technologies, industry best practices, and emerging trends. Regularly updating AI strategies based on evolving capabilities ensures sustained benefits and a competitive edge in the construction sector.

In summary, AI adoption strategies in construction firms involve thorough assessment, goal setting, collaboration, talent development, pilot projects, integration, data management, partnerships, performance measurement, and a commitment to continuous learning. A well-executed AI adoption strategy positions construction firms to harness the full potential of artificial intelligence, driving innovation and efficiency in the industry.

# 13.1 Change Management

Change management is a critical aspect of successfully adopting AI strategies in construction firms.

Implementing artificial intelligence involves transforming workflows, introducing new technologies, and altering the way teams operate. Here are key change management strategies for AI adoption in construction firms:

1. **Leadership Commitment:**

    - **Strategy:** Ensure strong leadership commitment to AI adoption.

    - **Action:** Leaders should communicate the vision, benefits, and long-term goals of AI adoption. Their commitment sets the tone for the entire organization and fosters a culture of openness to change.

2. **Clear Communication:**

   - **Strategy:** Develop a clear communication plan.

   - **Action:** Communicate the reasons behind AI adoption, its benefits, and the expected impact on job roles. Regularly update employees on the progress and address concerns transparently.

3. **Employee Involvement:**

   - **Strategy:** Involve employees in the AI adoption process.

   - **Action:** Solicit input from workers at various levels. Engage them in pilot programs and seek feedback. This involvement fosters a sense of ownership and reduces resistance to change.

4. **Training and Upskilling:**

   - **Strategy:** Provide comprehensive training programs.

   - **Action:** Invest in training to enhance employees' AI literacy. Equip them with the skills needed to operate and collaborate with AI technologies. This empowers the workforce and mitigates fears of job displacement.

5. **Pilot Programs:**

   - **Strategy:** Implement AI in pilot programs.

   - **Action:** Start with small-scale projects to test AI applications in real-world scenarios. Use pilot programs to identify challenges, gather feedback, and make adjustments before broader implementation.

6. **Gradual Implementation:**

   - **Strategy:** Implement AI gradually.

   - **Action:** Avoid abrupt changes. Gradual implementation allows employees to adapt incrementally, reducing the shock of a major shift. It also provides time for addressing challenges as they arise.

7. **Cross-Functional Collaboration:**

- **Strategy:** Encourage cross-functional collaboration.

- **Action:** Facilitate collaboration between different departments to share insights and expertise. This collaborative approach helps break down silos and ensures a holistic understanding of how AI impacts the organization.

8. **Change Champions:**

- **Strategy:** Identify and empower change champions.

- **Action:** Designate individuals who are enthusiastic about AI adoption as change champions. These individuals can inspire and guide their peers, sharing positive experiences and helping to overcome resistance.

9. **Flexible Work Environment:**

- **Strategy:** Promote a flexible work environment.

- **Action:** Support flexible work arrangements to accommodate changes in workflow and project management. This flexibility helps employees adapt to new ways of working with AI technologies.

10. **Performance Metrics and Feedback:**

- **Strategy:** Establish performance metrics and provide regular feedback.

- **Action:** Define clear metrics for assessing the impact of AI adoption. Regularly review progress and provide constructive feedback. Recognize and reward employees who excel in adapting to and utilizing AI tools.

11. **Addressing Concerns:**

- **Strategy:** Address concerns proactively.

- **Action:** Actively listen to employees' concerns and address them promptly. Whether related to job security, changes in job roles, or the learning curve, acknowledging and resolving concerns helps build trust.

12. **Continuous Improvement:**

- **Strategy:** Embrace a culture of continuous improvement.

- **Action:** Iterate on AI implementations based on feedback and lessons learned. Foster a culture where employees feel comfortable suggesting improvements and participating in the ongoing evolution of AI strategies.

### 13. Ethical Considerations:

- **Strategy:** Address ethical considerations.

- **Action:** Clearly communicate the ethical considerations of AI adoption, including privacy, bias, and transparency. Develop guidelines and policies to ensure responsible AI use in construction processes.

### 14. External Expertise:

- **Strategy:** Seek external expertise.

- **Action:** Collaborate with AI experts, consultants, and industry peers to gain insights into best practices. External expertise can provide valuable perspectives and guidance throughout the AI adoption journey.

Change management is a dynamic and ongoing process. By implementing these strategies, construction firms can navigate the challenges of AI adoption, build a culture of innovation, and ensure a smooth transition toward a more technologically advanced and collaborative future.

## 13.2 Organizational Culture

Creating an organizational culture conducive to AI adoption in construction firms is crucial for successful implementation and sustained innovation.

Here are key elements of an organizational culture that supports AI adoption strategies:

1. **Innovation and Openness:**

   - **Culture:** Foster a culture of innovation and openness to new technologies.

   - **Action:** Encourage employees to explore and suggest innovative solutions, including AI applications. Celebrate experimentation and learning from failures as part of the innovation process.

2. **Continuous Learning:**

- **Culture:** Embrace a culture of continuous learning.

- **Action:** Promote ongoing training and development programs to enhance employees' skills in AI and related technologies. Create a learning environment that supports adaptation to new tools and methodologies.

3. **Collaboration and Cross-Functional Teams:**

- **Culture:** Promote collaboration and cross-functional teamwork.

- **Action:** Break down silos between departments and encourage collaboration between construction, IT, and data science teams. Foster an environment where diverse expertise comes together to drive AI initiatives forward.

4. **Data-Driven Decision-Making:**

- **Culture:** Instill a culture of data-driven decision-making.

- **Action:** Encourage the use of data to inform decision-making processes. Promote the understanding that AI technologies rely on quality data, and decisions should be based on a combination of data insights and human expertise.

5. **Risk-Taking and Experimentation:**

- **Culture:** Encourage a culture of calculated risk-taking and experimentation.

- **Action:** Provide a safe space for employees to experiment with AI applications. Acknowledge that not every experiment will succeed, but each failure is an opportunity to learn and improve.

6. **Leadership Support and Vision:**

- **Culture:** Ensure leadership support and a clear vision for AI adoption.

- **Action:** Leaders should communicate the strategic importance of AI adoption, outlining the benefits for the organization and its employees. Leadership support sets the tone for the entire organization.

7. **Agility and Adaptability:**

   - **Culture:** Foster agility and adaptability.

   - **Action:** Build an organizational culture that can quickly adapt to changes in technology and market demands. Agility is crucial for responding to challenges and seizing opportunities in the rapidly evolving landscape of AI.

8. **Ethical Considerations:**

   - **Culture:** Prioritize ethical considerations in AI adoption.

   - **Action:** Establish and communicate ethical guidelines for the use of AI technologies. Create awareness among employees about the responsible and ethical use of AI to build trust within the organization and with external stakeholders.

9. **Inclusive and Diverse Environment:**

   - **Culture:** Cultivate an inclusive and diverse workplace.

   - **Action:** Recognize the value of diverse perspectives in AI development and application. An inclusive environment encourages collaboration and ensures that AI technologies consider a broad range of perspectives and potential biases.

10. **Communication and Transparency:**

    - **Culture:** Prioritize open communication and transparency.

    - **Action:** Keep employees informed about the AI adoption process, including goals, challenges, and progress. Address concerns transparently and maintain open channels for feedback.

11. **Recognition and Reward System:**

    - **Culture:** Implement a recognition and reward system for AI contributions.

- **Action:** Acknowledge and reward employees who actively contribute to AI initiatives. Recognition can be both financial and non-financial, fostering a sense of accomplishment and motivation.

## 12. Customer-Centric Focus:

- **Culture:** Instill a customer-centric mindset.

- **Action:** Ensure that AI applications align with customer needs and expectations. A customer-focused culture drives the development of AI solutions that enhance client satisfaction and contribute to business success.

## 13. Security Awareness:

- **Culture:** Promote a culture of cybersecurity and data protection.

- **Action:** Emphasize the importance of data security and privacy. Train employees to be vigilant against potential cybersecurity threats associated with AI technologies.

## 14. Balancing AI and Human Capabilities:

- **Culture:** Emphasize the complementary nature of AI and human capabilities.

- **Action:** Communicate that AI technologies are tools to augment human abilities, not replace them. Encourage a culture that values the unique strengths that both humans and AI bring to the table.

Building a culture that aligns with AI adoption strategies is an ongoing effort that requires commitment from leadership, active involvement from employees, and a shared vision for the future of the organization. By embedding these cultural elements, construction firms can create an environment where AI adoption is not just a technological change but a holistic transformation embraced by the entire organization.

## 13.3 Cost-Benefit Analysis

A comprehensive cost-benefit analysis (CBA) is essential for construction firms considering the adoption of AI strategies.

This analysis helps evaluate the financial feasibility and potential returns on investment associated with implementing AI technologies. Here are key components to consider in a cost-benefit analysis for AI adoption in construction firms:

1. **Costs of AI Adoption:**

   i. **Initial Implementation Costs:**

- Cost: Expenses associated with purchasing and implementing AI technologies, including software, hardware, and consulting services.

- Breakdown: Detailed breakdown of upfront costs for AI adoption.

ii. **Training and Skill Development:**

- Cost: Investments in training programs for employees to acquire the necessary skills for AI collaboration.

- Breakdown: Costs associated with training materials, courses, and potential productivity losses during the learning curve.

iii. **Integration with Existing Systems:**

- Cost: Expenses related to integrating AI systems with existing IT infrastructure and software.

- Breakdown: Costs for customization, software development, and potential disruptions during integration.

iv. **Maintenance and Upkeep:**

- Cost: Ongoing expenses for maintaining and updating AI systems to ensure optimal performance.

- Breakdown: Costs associated with regular maintenance, software updates, and support services.

v. **Data Security and Compliance:**

- Cost: Investments in cybersecurity measures, data protection, and compliance with relevant regulations.

- Breakdown: Costs for cybersecurity tools, data encryption, and legal compliance efforts.

vi. **Change Management:**

- Cost: Expenses related to change management initiatives, including communication, employee training, and addressing resistance.

- Breakdown: Costs associated with change management programs, workshops, and employee engagement activities.

vii. **Monitoring and Evaluation:**

- Cost: Resources allocated for monitoring the performance of AI systems and evaluating their impact.

- Breakdown: Costs for analytics tools, data monitoring, and evaluation processes.

2. **Benefits of AI Adoption:**

i. **Increased Efficiency:**

- Benefit: AI technologies can automate repetitive tasks, optimize processes, and enhance overall project efficiency.

- Measurement: Evaluate the time saved and increased productivity resulting from the implementation of AI.

ii. **Improved Accuracy and Quality:**

- Benefit: AI applications can enhance accuracy in tasks such as design, planning, and quality control.

- Measurement: Assess the reduction in errors, rework, and the overall improvement in project outcomes.

iii. **Cost Savings:**

- Benefit: Automation and optimization can lead to reduced labor costs, resource wastage, and operational expenses.

- Measurement: Quantify the direct cost savings associated with AI implementation across various aspects of construction processes.

iv. **Enhanced Safety:**

- Benefit: AI technologies can contribute to improved safety through monitoring, predictive maintenance, and risk analysis.

- Measurement: Evaluate the reduction in accidents, insurance costs, and associated expenses related to safety incidents.

v. **Faster Decision-Making:**

- Benefit: AI-driven analytics and decision support systems can expedite decision-making processes.

- Measurement: Measure the time saved in decision-making and the impact on project timelines.

vi. **Optimized Resource Allocation:**

- Benefit: AI can assist in better resource allocation, including materials, equipment, and manpower.

- Measurement: Assess the efficiency gains and cost reductions achieved through optimized resource allocation.

vii. **Predictive Maintenance:**

- Benefit: AI applications can predict equipment failures, reducing downtime and maintenance costs.

- Measurement: Evaluate the decrease in unplanned maintenance activities and associated costs.

viii. **Competitive Advantage:**

- Benefit: Early adoption of AI can provide a competitive edge in bidding, project delivery, and client satisfaction.

- Measurement: Assess the impact on market share, client acquisition, and business growth.

3. **Calculating Return on Investment (ROI):**

Net Benefit = Total Benefits - Total Costs

Calculate ROI:

$$ROI = \left(\frac{NetBenefit}{TotalCosts}\right) \times 100$$

4. **Additional Considerations:**

i. **Risk Assessment:**

- Evaluate potential risks associated with AI adoption, including technical challenges, regulatory changes, and market uncertainties.

ii.   **Scenario Analysis:**

   o   Conduct scenario analyses to understand the impact of different assumptions and variables on the cost-benefit outcomes.

iii.  **Long-Term Strategic Alignment:**

   o   Consider the long-term strategic alignment of AI adoption with the organization's goals and vision.

iv.   **Non-Financial Considerations:**

   o   Include non-financial considerations such as improved customer satisfaction, brand reputation, and employee morale.

v.    **Benchmarking:**

   o   Benchmark the construction firm's AI adoption against industry standards and competitors to gain insights into potential performance improvements.

A thorough cost-benefit analysis will provide construction firms with a holistic understanding of the financial implications and potential gains associated with the adoption of AI strategies. This analysis serves as a crucial decision-making tool for organizations considering the integration of AI technologies into their construction processes.

# 13.4 Risk Mitigation

Implementing AI adoption strategies in construction firms involves certain risks, and effective risk mitigation is essential to ensure the successful integration of AI technologies.

Here are key risk factors and corresponding mitigation strategies:

7. **Technical Risks:**

    i. **System Integration Issues:**

        • Risk: Difficulty integrating AI systems with existing construction management software and processes.

        • Mitigation:

    o  Conduct thorough compatibility assessments.

    o  Choose AI solutions with flexible integration capabilities.

    o  Pilot test integration in a controlled environment.

## ii. Data Quality and Availability:

- Risk: Insufficient or poor-quality data for training AI models.

- Mitigation:

    o  Invest in data quality assessments and cleansing.

    o  Implement data governance practices.

    o  Ensure data sources are reliable and up-to-date.

## iii. Algorithmic Bias:

- Risk: Bias in AI algorithms leading to unfair decision-making.

- Mitigation:

    o  Regularly audit algorithms for bias.

    o  Diversify training datasets to avoid biases.

    o  Implement fairness-aware machine learning practices.

## 8. Operational Risks:

### i. Workforce Resistance:

- Risk: Employee resistance to AI adoption due to fear of job displacement or unfamiliarity.

- Mitigation:

    o  Develop comprehensive training programs.

    o  Communicate the benefits of AI for job augmentation.

    o  Involve employees in the AI planning and implementation process.

ii.    **Lack of Skills:**

- Risk: Inadequate skills among the workforce to effectively use and collaborate with AI technologies.

- Mitigation:
    - Invest in training programs for employees.
    - Encourage continuous learning and upskilling.
    - Hire or consult with AI specialists.

iii.    **Change Management Challenges:**

- Risk: Ineffective change management leading to disruptions and productivity losses.

- Mitigation:
    - Develop a robust change management plan.
    - Communicate the vision and benefits of AI adoption.
    - Provide ongoing support and address concerns proactively.

iv.    **Security and Compliance Risks:**

- Data Security and Privacy:

- Risk: Unauthorized access, data breaches, or privacy violations.

- Mitigation:
    - Implement robust cybersecurity measures.
    - Encrypt sensitive data.
    - Comply with relevant data protection regulations.

v.    **Regulatory Compliance:**

- Risk: Non-compliance with industry regulations and standards.

- Mitigation:

- o Stay informed about relevant regulations.
- o Implement compliance measures in AI development and usage.
- o Engage legal experts to ensure adherence to regulations.

## 9. Financial Risks:

### i. Unforeseen Costs:

- Risk: Unexpected costs associated with AI adoption.
- Mitigation:
  - o Conduct a comprehensive cost-benefit analysis.
  - o Create contingency budgets for unforeseen expenses.
  - o Regularly review and update financial projections.

### ii. ROI Not Achieved:

- Risk: Failure to achieve the expected return on investment.
- Mitigation:
  - o Set realistic and measurable goals.
  - o Continuously monitor and evaluate the performance of AI implementations.
  - o Be prepared to adjust strategies based on ongoing assessments.

## 10. External Risks:

### i. Vendor Reliability:

- Risk: Dependence on AI vendors with potential reliability issues.
- Mitigation:
  - o Thoroughly vet AI vendors.

- o Establish service-level agreements (SLAs) with clear expectations.

- o Diversify vendors when possible.

ii. **Rapid Technological Changes:**

- Risk: Rapid technological advancements making adopted AI technologies obsolete.

- Mitigation:

  - o Stay informed about emerging technologies.

  - o Choose scalable and adaptable AI solutions.

  - o Plan for regular updates and upgrades.

## 11. Ethical and Reputation Risks:

i. **Ethical Concerns:**

- Risk: Ethical considerations, such as biased decision-making, causing reputational damage.

- Mitigation:

  - o Implement ethical guidelines for AI usage.

  - o Regularly audit AI algorithms for bias.

  - o Communicate transparently about ethical practices.

ii. **Stakeholder Perception:**

- Risk: Negative perception among stakeholders regarding AI adoption.

- Mitigation:

  - o Communicate the benefits of AI in construction projects.

  - o Engage with stakeholders to address concerns and gather feedback.

  - o Showcase successful AI implementations and positive outcomes.

**12. General Risk Mitigation Strategies:**

i.   **Thorough Planning and Assessment:**

- Conduct thorough risk assessments before implementing AI strategies.

- Develop detailed plans for AI adoption, including risk mitigation strategies.

ii.  **Pilot Testing:**

- Implement pilot programs before full-scale deployment to identify and address potential issues.

iii. **Regular Monitoring and Evaluation:**

- Continuously monitor the performance of AI systems and evaluate their impact on construction processes.

iv.  **Legal and Regulatory Compliance:**

- Stay abreast of legal and regulatory requirements related to AI in construction and ensure compliance.

v.   **Cross-Functional Collaboration:**

- Foster collaboration between IT, data science, and construction teams to collectively address challenges and risks.

vi.  **Ethical Guidelines:**

- Establish and communicate clear ethical guidelines for the development and use of AI technologies.

vii. **Crisis Communication Plan:**

- Develop a crisis communication plan to address unforeseen challenges and maintain transparency.

viii. **Diversification of AI Applications:**

- Diversify AI applications across different aspects of construction to reduce dependence on a single technology.

ix.  **Employee Engagement:**

- Actively involve employees in the AI adoption process, addressing concerns and seeking their input.

x. **Scenario Planning:**

- Conduct scenario analyses to anticipate potential risks and develop corresponding mitigation strategies.

By proactively identifying and mitigating these risks, construction firms can navigate the challenges associated with AI adoption and maximize the benefits of integrating artificial intelligence into their operations.

## 13.5 Success Stories and Lessons Learned

There were several success stories and lessons learned from AI adoption strategies in construction firms.

Here are some illustrative examples and insights:

### A. Success Stories:

1. **Doxel:**

   - **Overview:** Doxel is an AI startup that uses robots and AI to monitor construction sites in real-time.

   - **Success:** Doxel's AI platform helped a large construction project in California stay on schedule and within budget by providing accurate progress tracking, identifying potential issues early, and optimizing resource allocation.

- **Lesson Learned:** Real-time monitoring and AI-driven insights can significantly enhance project management and decision-making.

2. **Autodesk Construction Solutions:**

- **Overview:** Autodesk's BIM 360 and other construction management solutions incorporate AI to streamline workflows.

- **Success:** AI-powered tools help construction teams manage documents, collaborate more efficiently, and optimize project schedules.

- **Lesson Learned:** Integration of AI into construction management software can improve project coordination and communication.

3. **Brickbots:**

- **Overview:** Brickbots is a construction robotics company that uses AI and robotics for bricklaying.

- **Success:** The use of robotic bricklayers equipped with AI vision systems has increased the speed and precision of bricklaying processes, reducing manual labor and improving accuracy.

- **Lesson Learned:** Robotics and AI can be applied to specific construction tasks, augmenting human capabilities and improving efficiency.

4. **Trimble Construction Software:**

- **Overview:** Trimble offers construction software solutions, including AI-powered tools for project estimation and planning.

- **Success:** AI algorithms analyze historical project data to provide accurate cost estimates and optimize project schedules, leading to better-informed decision-making.

- **Lesson Learned:** AI-driven analytics can enhance project planning and estimation accuracy.

B. **Lessons Learned:**

i. **Start with Pilot Projects:**

- Begin AI adoption with smaller, well-defined pilot projects to test feasibility, identify challenges, and refine strategies before scaling up.

ii. **Employee Training and Involvement:**

- Invest in comprehensive training programs to equip employees with the skills needed to collaborate with AI technologies. Involving employees in the adoption process fosters a positive attitude toward AI.

iii. **Data Quality is Key:**

- The success of AI applications relies on the quality of data. Ensure that data used for training AI models is accurate, reliable, and representative of the construction environment.

iv. **Collaboration Platforms Matter:**

- Implement collaborative platforms that facilitate communication and coordination among project stakeholders. These platforms, powered by AI, can enhance project visibility and decision-making.

v. **Iterative Improvement:**

- Embrace an iterative approach to AI implementation. Continuously monitor performance, gather feedback, and make adjustments to improve the effectiveness of AI applications over time.

vi. **Customization for Industry Specifics:**

- AI solutions in construction may need customization to address industry-specific challenges. Tailoring AI applications to the unique requirements of the construction sector enhances their relevance and impact.

vii. **Addressing Ethical Considerations:**

- Proactively address ethical considerations associated with AI, such as bias and transparency. Establish guidelines to ensure responsible and ethical use of AI technologies in construction.

viii. **Risk Management is Crucial:**

- Identify potential risks and develop robust risk mitigation strategies. Proactive risk management is essential for minimizing disruptions and ensuring the success of AI adoption.

ix. **Scalability and Integration:**

- Choose AI solutions that are scalable and can seamlessly integrate with existing construction management systems. This facilitates a smoother transition and long-term sustainability.

x. **Leadership Commitment:**

- Leadership commitment is critical for the success of AI adoption. Clear vision, support, and communication from leadership create a positive organizational culture around technology integration.

xi. **Real-Time Monitoring Pays Off:**

- Real-time monitoring using AI technologies provides actionable insights for timely decision-making. This can lead to improved project outcomes and cost savings.

xii. **Customer-Centric Solutions:**

- AI adoption should align with customer needs and expectations. Solutions that enhance customer satisfaction and contribute to positive client experiences are likely to be more successful.

These lessons highlight the importance of a strategic and thoughtful approach to AI adoption in construction firms. Success is often driven by a combination of technology, organizational culture, and a commitment to ongoing improvement.

# 13.6 Real-world Examples / Use Cases

1. **AI Adoption Strategies in Construction Firms:**

    i. **Data-Driven Decision-Making (Company: Turner Construction):**

    o Turner Construction implements AI adoption strategies focused on data-driven decision-making. By leveraging AI analytics, Turner enhances project management, risk assessment, and overall decision-making processes in construction.

    ii. **Predictive Maintenance Implementation (Company: Skanska):**

    o Skanska adopts AI in construction for predictive maintenance. Utilizing machine learning algorithms, Skanska optimizes the maintenance schedules of construction equipment, reducing downtime and improving operational efficiency.

2. **Change Management for AI Adoption in Construction Firms:**

    iii. **Change Management Consulting (Company: Change Leadership):**

    o Change Leadership provides change management consulting services for AI adoption in construction firms. Their expertise assists organizations in transitioning smoothly to AI technologies, ensuring effective integration and employee acceptance.

    iv. **Internal Training Programs (Company: DPR Construction):**

    o DPR Construction focuses on internal training programs for change management in AI adoption. By educating employees on the benefits and applications of AI, DPR ensures a culture of adaptability and innovation.

3. **Organizational Culture for AI Adoption in Construction Firms:**

    v. **Innovation Culture (Company: Katerra):**

- o Katerra cultivates an innovation culture to support AI adoption in construction. By fostering a mindset of continuous improvement and openness to technological advancements, Katerra aligns its organizational culture with AI integration.

  vi. **Collaborative Approach (Company: McCarthy Building Companies):**

- o McCarthy Building Companies emphasizes a collaborative approach to organizational culture for AI adoption. This involves encouraging teamwork, knowledge sharing, and cross-functional collaboration to facilitate a smooth transition to AI technologies.

4. **Cost-Benefit Analysis for AI Adoption in Construction Firms:**

  vii. **Cost Savings through AI Analytics (Company: Mortenson):**

- o Mortenson conducts a cost-benefit analysis for AI adoption, particularly in analytics. By realizing cost savings through improved efficiency and resource allocation, Mortenson justifies the investment in AI technologies.

  viii. **ROI Assessment for Construction Automation (Company: Balfour Beatty):**

- o Balfour Beatty performs a comprehensive ROI assessment for AI adoption, particularly in construction automation. By quantifying the returns on investment, Balfour Beatty ensures strategic decision-making regarding the adoption of AI in construction.

5. **Risk Mitigation in AI Adoption in Construction Firms:**

  ix. **Cybersecurity Measures (Company: Clark Construction):**

- o Clark Construction prioritizes risk mitigation in AI adoption by implementing robust cybersecurity measures. Ensuring the security of AI systems protects sensitive construction data from potential threats and vulnerabilities.

  x. **Pilot Project Testing (Company: AECOM):**

○ AECOM adopts a risk mitigation strategy through pilot project testing. By initially implementing AI technologies on a smaller scale, AECOM assesses performance, identifies potential challenges, and mitigates risks before broader adoption.

## 13.7 Chapter Summary: Key Points

1. AI adoption in construction firms involves systematic integration for project planning, execution, and management, starting with a comprehensive assessment of current workflows.

2. Clear objectives, use cases, and measurable goals are crucial, with robust data management, cross-functional collaboration, and pilot projects playing key roles in successful AI implementation.

3. Fostering a culture of collaboration and communication, investing in talent and technology, and ensuring seamless integration with existing systems are essential steps in AI adoption.

4. Data management, including establishing robust governance practices, is critical for the success of AI applications in construction.

5. Building partnerships with AI solution providers and technology experts accelerates adoption by facilitating knowledge transfer and access to cutting-edge technologies.

6. Measuring success through key performance indicators (KPIs) aligned with goals and continuous learning and adaptation are integral for staying at the forefront of AI innovation.

7. Change management is crucial, requiring strong leadership commitment, clear communication, employee involvement, and comprehensive training programs.

8. Organizational culture supporting AI adoption involves fostering innovation, continuous learning, collaboration, and a customer-centric focus.

9. Key elements include promoting data-driven decision-making, risk-taking, leadership support, agility, ethical considerations, and a diverse and inclusive environment.

10. Cost-benefit analysis for AI adoption involves assessing initial costs, training, integration, maintenance, and data security against benefits like increased efficiency, accuracy, cost savings, and competitive advantage.

11. Calculating return on investment (ROI) helps determine the net benefit and includes considerations for risk assessment, scenario analysis, and long-term strategic alignment.

12. Risk mitigation strategies address technical, operational, security, financial, and external risks, emphasizing thorough planning, pilot testing, and continuous monitoring.

13. Success stories include companies like Doxel, Autodesk, Brickbots, and Trimble, showcasing real-time monitoring, robotics, and AI-driven analytics in construction.

14. Lessons learned stress the importance of starting with pilot projects, employee training, data quality, collaboration platforms, iterative improvement, ethical considerations, and risk management.

15. Scalability, integration, leadership commitment, real-time monitoring, and customer-centric solutions are key factors contributing to the success of AI adoption in construction firms.

## 13.8 Concept Check: Q&A Sessions

1. What is the initial step for construction firms in the AI adoption process, and why is it crucial?

2. How can construction firms foster collaboration during AI adoption, and why is it emphasized?

3. Why is the integration of AI with existing workflows in construction firms recommended to begin with pilot projects?

4. What role does leadership commitment play in successful AI adoption, and how can leaders contribute?

5. How can construction firms address employee resistance to AI adoption?

6. What are the key cultural elements necessary for supporting AI adoption in construction firms?

7. Why is a comprehensive cost-benefit analysis (CBA) essential for construction firms considering AI adoption?

8. How can construction firms mitigate technical risks associated with AI adoption, particularly in system integration?

9. What role does ethical considerations play in AI adoption, and how can firms address ethical concerns?

10. Why is scenario analysis included in the cost-benefit analysis for AI adoption in construction firms?

# 14. Future Innovations in AI

The future of AI in civil engineering holds promising innovations that are poised to revolutionize the industry.

One notable area is the advancement of Generative Design, where AI algorithms will play a pivotal role in generating and optimizing design alternatives. This innovation enables engineers to explore a multitude of design possibilities, considering various constraints and requirements, ultimately leading to more efficient and creative solutions.

Predictive maintenance powered by AI is expected to evolve further, enhancing the lifespan and reliability of infrastructure. Machine learning models will become more sophisticated in analyzing data from sensors to predict equipment failures, optimizing maintenance schedules, and reducing downtime. The future of civil engineering is set to be revolutionized by advancements in artificial intelligence (AI), with the following key innovations on the horizon.

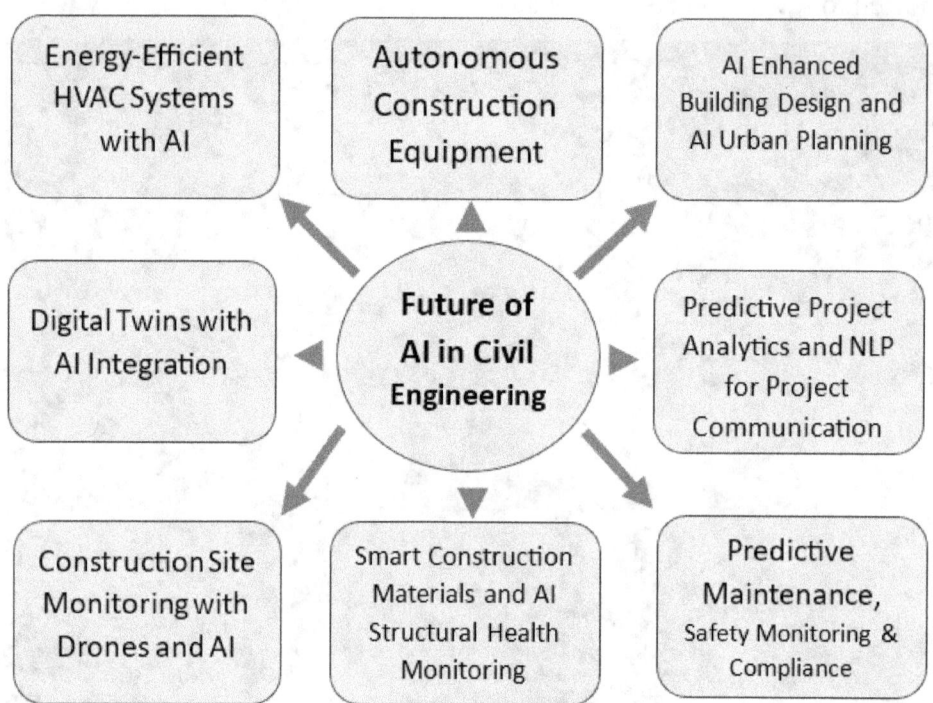

The future of civil engineering is set to be revolutionized by advancements in artificial intelligence (AI), with several key innovations on the horizon. This includes the integration of AI into construction machinery, paving the way for autonomous operation of equipment like excavators and cranes, optimizing efficiency and safety. AI-powered generative design tools are expected to further evolve, assisting architects and engineers in creating optimized building designs that prioritize sustainability and energy efficiency. Enhanced predictive analytics using AI algorithms will revolutionize project planning, enabling more accurate forecasting of timelines, costs, and potential risks. The development of smart construction materials with self-monitoring capabilities, such as adaptive and self-repairing properties, is anticipated. Additionally, the combined use of drones and AI for real-time construction site monitoring, AI-driven structural health monitoring, and the integration of AI with digital twin technology for realistic simulations represent key advancements. AI is also anticipated to play a significant role in urban planning, optimizing city layouts, transportation systems, and infrastructure development. Further innovations include the integration of natural language processing into project communication systems and the continued development of AI-driven HVAC systems for energy-efficient building operation. These innovations collectively underscore AI's transformative impact on civil engineering, promising increased efficiency, sustainability, and overall advancements in construction and infrastructure development.

AI's role in autonomous construction machinery is a frontier that promises increased efficiency and safety. As technology advances, construction equipment equipped with AI algorithms will gain greater autonomy, enabling them to operate more intelligently, make real-time decisions, and collaborate seamlessly with human workers.

The integration of AI with Building Information Modeling (BIM) is set to transform project planning and management. AI algorithms will analyze vast datasets from BIM models, providing insights into potential issues, optimizing project schedules, and contributing to more informed decision-making throughout the construction lifecycle.

Real-time collaboration in construction projects will see advancements with AI-driven tools facilitating communication and coordination among project stakeholders. Enhanced virtual collaboration spaces, powered by AI, will enable teams to interact, share insights, and make collective decisions irrespective of geographical locations.

The development of AI-driven safety monitoring systems is crucial for improving workplace safety in construction. AI algorithms will analyze data from wearable devices, video feeds, and other sensors to detect potential safety hazards, predict risky behaviors, and provide real-time alerts to prevent accidents.

AI's role in materials science will lead to innovations in the development of advanced construction materials. Machine learning models will analyze material properties, environmental conditions, and performance data to design materials with enhanced durability, sustainability, and structural integrity.

The application of AI in risk assessment and mitigation will become more sophisticated. Machine learning models will assess project risks by analyzing historical data, project complexities, and external factors, providing project managers with actionable insights to mitigate potential issues proactively.

In the field of environmental sustainability, AI will contribute to optimizing energy-efficient building designs. Machine learning algorithms will analyze environmental data, energy consumption patterns, and local climate conditions to inform architects and engineers on sustainable design choices that minimize environmental impact.

AI's integration with robotics in construction will continue to evolve, leading to the development of more advanced robotic systems with enhanced autonomy and adaptability. These AI-driven robots will be capable of performing a wider range of tasks, from intricate assembly work to complex construction activities, improving overall efficiency and precision in construction processes.

In conclusion, the future innovations in AI for civil engineering encompass a wide spectrum of applications, from generative design and predictive maintenance to autonomous machinery, enhanced collaboration tools, safety monitoring, advanced materials, risk assessment, sustainable design, and evolving robotics. These innovations collectively represent a transformative trajectory for the civil engineering industry, contributing to increased efficiency, sustainability, and safety in construction practices.

## 14.1 Emerging Technologies

There are several emerging technologies were expected to shape future innovations in AI for civil engineering.

Here are some key emerging technologies that were anticipated to influence the future of AI in civil engineering:

1. **Generative Design:**
   o **Overview:** Generative design involves using AI algorithms to explore numerous design possibilities and automatically generate optimized solutions based on specified constraints and objectives.

- **Impact on Civil Engineering:** Generative design can aid in creating innovative and efficient designs for structures, infrastructure, and urban planning, optimizing material usage and overall project performance.

2. **Digital Twins:**

   - **Overview:** Digital twins involve creating virtual replicas of physical structures or systems, leveraging real-time data for monitoring, analysis, and simulation.

   - **Impact on Civil Engineering:** AI-driven digital twins allow for continuous monitoring, predictive maintenance, and performance optimization of infrastructure projects, enhancing decision-making and project lifecycle management.

3. **Edge Computing:**

   - **Overview:** Edge computing involves processing data near the source of generation rather than relying on centralized cloud servers.

   - **Impact on Civil Engineering:** In civil engineering applications, edge computing can enable real-time data analysis, facilitating quick decision-making on construction sites and infrastructure monitoring.

4. **Augmented Reality (AR) and Virtual Reality (VR):**

   - **Overview:** AR overlays digital information onto the real-world environment, while VR creates immersive virtual experiences.

   - **Impact on Civil Engineering:** AR and VR can be used for visualizing construction projects, conducting virtual inspections, and providing training simulations for construction personnel.

5. **Blockchain:**

   - **Overview:** Blockchain is a decentralized and secure ledger technology that records and verifies transactions across a network of computers.

- o **Impact on Civil Engineering:** In civil engineering, blockchain can enhance transparency, traceability, and security in project management, supply chain, and payment systems.

6. **Internet of Things (IoT):**

- o **Overview:** IoT involves connecting physical devices and sensors to the internet for data collection and communication.

- o **Impact on Civil Engineering:** IoT applications in civil engineering include smart infrastructure with sensors for structural health monitoring, environmental sensing, and real-time data collection for better decision-making.

7. **5G Technology:**

- o **Overview:** 5G is the fifth generation of mobile networks, providing high-speed, low-latency communication.

- o **Impact on Civil Engineering:** 5G facilitates real-time communication and data exchange, enabling efficient connectivity for IoT devices, autonomous construction equipment, and collaborative project management.

8. **Natural Language Processing (NLP):**

- o **Overview:** NLP involves the interaction between computers and human language, enabling machines to understand, interpret, and generate human-like text.

- o **Impact on Civil Engineering:** NLP can be applied for better communication between project stakeholders, automated documentation, and analysis of textual data related to construction projects.

9. **Robotics and Autonomous Construction Equipment:**

- o **Overview:** Robotics involves the design and operation of robots, and autonomous construction equipment can perform tasks without direct human intervention.

- o **Impact on Civil Engineering:** AI-powered robotics and autonomous equipment can enhance efficiency, safety, and precision in construction processes, including tasks like excavation and material handling.

### 10. Machine Learning Algorithms for Predictive Analytics:

- **Overview:** Machine learning algorithms analyze data patterns to make predictions and decisions without explicit programming.

- **Impact on Civil Engineering:** Predictive analytics using machine learning can help anticipate project risks, optimize scheduling, and improve resource allocation based on historical data and project-specific parameters.

### 11. Quantum Computing:

- **Overview:** Quantum computing leverages the principles of quantum mechanics to perform complex computations.

- **Impact on Civil Engineering:** While still in the early stages, quantum computing has the potential to significantly accelerate complex simulations, optimization problems, and data analysis in civil engineering applications.

These emerging technologies, when integrated with AI, are expected to drive transformative changes in civil engineering, offering improved efficiency, sustainability, and innovation in construction practices. It's crucial for professionals in the field to stay updated on these developments to leverage the full potential of AI-driven advancements in civil engineering.

## 14.2 Research Frontiers

There are several research frontiers were actively explored for future innovations in AI for civil engineering.

The field is dynamic, and new research areas may have emerged since then. Here are some research frontiers that were gaining attention:

1. **Explainable AI (XAI) for Civil Engineering:**
   - Research Focus: Developing AI models that provide transparent and interpretable results in civil engineering applications. Understanding and explaining the decision-making processes of AI systems is crucial for gaining trust and ensuring accountability.

2. **Hybrid AI Systems:**
   - Research Focus: Investigating the integration of various AI techniques, including machine learning, expert systems, and

optimization algorithms, to create hybrid AI systems. These systems aim to leverage the strengths of different approaches for improved performance in civil engineering tasks.

3. **AI for Sustainable Design and Construction:**

   - Research Focus: Applying AI to optimize sustainable practices in the design and construction of buildings and infrastructure. This includes energy-efficient designs, material selection for reduced environmental impact, and life cycle assessment using AI algorithms.

4. **Resilience and Risk Assessment with AI:**

   - Research Focus: Using AI to assess and enhance the resilience of civil infrastructure to natural disasters and climate change. Research explores AI-based risk assessment models, early warning systems, and decision support tools for resilient infrastructure planning.

5. **Human-AI Collaboration in Construction:**

   - Research Focus: Investigating ways to enhance collaboration between AI systems and human workers in construction processes. This includes developing AI tools that assist and augment human capabilities, taking into account the unique skills and expertise of construction professionals.

6. **Edge AI for Real-Time Monitoring:**

   - Research Focus: Exploring the potential of edge computing and AI for real-time monitoring of construction sites. Research aims to develop efficient edge AI systems capable of processing and analyzing data locally, reducing latency and improving responsiveness in monitoring applications.

7. **Digital Twins and AI Simulation:**

   - Research Focus: Advancing the integration of AI with digital twins for more realistic and dynamic simulations of construction projects. Research explores how AI can enhance the accuracy of digital twin models, enabling better predictions and decision-making.

8. **AI-Driven Predictive Maintenance:**

- Research Focus: Investigating AI applications for predictive maintenance of infrastructure. Research aims to develop AI models that can predict the maintenance needs of structures, bridges, and other civil assets, optimizing maintenance schedules and reducing downtime.

9. **Autonomous Construction Equipment and Robotics:**

- Research Focus: Advancing the capabilities of autonomous construction equipment through AI and robotics. Research explores the development of AI-driven algorithms for navigation, coordination, and decision-making in autonomous construction processes.

10. **Natural Language Processing (NLP) in Project Management:**

- Research Focus: Exploring the use of NLP in improving communication and collaboration in construction project management. Research investigates AI-powered NLP tools for analyzing and understanding project-related documents, facilitating effective communication among project stakeholders.

11. **Quantum Computing for Complex Simulations:**

- Research Focus: Investigating the potential of quantum computing for solving complex simulation and optimization problems in civil engineering. Research explores how quantum algorithms can accelerate computations related to structural analysis, material design, and optimization.

12. **Robustness and Security of AI Systems:**

- Research Focus: Addressing the robustness and security challenges associated with AI applications in civil engineering. Research explores methods to enhance the resilience of AI models to adversarial attacks and ensure the security of AI-driven infrastructure systems.

Researchers are actively exploring these frontiers to push the boundaries of what AI can achieve in the field of civil engineering. As technology continues to advance, these research areas are likely to contribute to the development of innovative solutions, improving the efficiency, sustainability, and resilience of civil infrastructure.

## 14.3 Industry Collaborations

Collaborations between the academic and industrial sectors are essential for driving future innovations in AI for civil engineering.

Such partnerships leverage the expertise of researchers, the practical insights of industry professionals, and the resources of both sectors. Here are some examples and potential areas of collaboration:

1. **Research Institutions and Construction Companies:**

   - Collaboration Focus: Joint research projects between universities and construction companies to explore AI applications in construction processes. This includes optimizing project scheduling, resource allocation, and the development of AI-driven construction technologies.

2. **Technology Companies and Engineering Firms:**

   - Collaboration Focus: Partnerships between technology companies specializing in AI solutions and engineering firms.

These collaborations aim to integrate AI technologies into engineering workflows, enhancing design optimization, structural analysis, and project management.

3. **Government Agencies and AI Startups:**

   - Collaboration Focus: Collaborations between government agencies responsible for infrastructure development and AI startups. These partnerships may focus on implementing AI for smart city planning, traffic management, and the development of sustainable infrastructure.

4. **Materials Science Researchers and Construction Material Manufacturers:**

   - Collaboration Focus: Joint efforts between materials science researchers and construction material manufacturers to apply AI for the development of advanced, sustainable materials. This includes AI-driven research on new construction materials with enhanced durability and environmental performance.

5. **Real Estate Developers and AI Innovators:**

   - Collaboration Focus: Collaborations between real estate developers and AI innovators to apply AI in real estate planning, design, and construction. This includes AI-powered tools for site selection, predictive modeling for property valuation, and smart building technologies.

6. **Infrastructure Maintenance Providers and AI Specialists:**

   - Collaboration Focus: Collaborations between companies specializing in infrastructure maintenance and AI specialists. These partnerships aim to develop AI applications for predictive maintenance, condition monitoring, and rehabilitation planning of existing civil infrastructure.

7. **Geospatial Technology Companies and Surveying Firms:**

   - Collaboration Focus: Partnerships between geospatial technology companies and surveying firms to integrate AI into geospatial data analysis. This includes AI-driven solutions for land surveying, terrain analysis, and mapping for construction and infrastructure projects.

8. **Construction Robotics Companies and Automation Experts:**

- Collaboration Focus: Collaboration between construction robotics companies and experts in automation and control systems. This collaboration aims to advance the integration of AI-driven robotics in construction processes, including autonomous construction equipment and robotic assembly.

9. **Smart Infrastructure Initiatives and IoT Solution Providers:**

- Collaboration Focus: Partnerships between organizations leading smart infrastructure initiatives and companies specializing in IoT solutions. These collaborations focus on implementing AI and IoT technologies for real-time monitoring, data analytics, and optimization of smart infrastructure systems.

10. **Construction Software Developers and AI Researchers:**

- Collaboration Focus: Collaborations between companies developing construction management software and AI researchers. These partnerships aim to enhance project management software with AI capabilities, including predictive analytics, risk assessment, and decision support.

11. **Cross-Industry Collaboration:**

- Collaboration Focus: Encouraging collaboration between the civil engineering sector and other industries, such as healthcare, aerospace, or automotive. Cross-industry collaborations can bring fresh perspectives and innovative AI solutions to address common challenges in diverse fields.

These collaborative efforts contribute to the development and deployment of AI technologies that address real-world challenges in civil engineering. The synergies created through industry-academic partnerships foster a collaborative ecosystem where research findings are translated into practical applications, benefiting both the academic community and the industry. It's essential for stakeholders to actively seek and promote such collaborations to accelerate the pace of innovation in AI for civil engineering.

## 14.4 Interdisciplinary Applications

Interdisciplinary applications of AI in civil engineering involve leveraging insights, methodologies, and technologies from various fields to address complex challenges and drive innovation.

Here are some key areas of interdisciplinary applications:

1. **Computational Biology and Structural Engineering:**

   - Application: Drawing inspiration from biological systems, researchers explore AI algorithms to optimize the structural design of buildings and bridges. This includes mimicking biological growth processes for more efficient and sustainable designs.

2. **Climate Science and Infrastructure Resilience:**

   - Application: Integrating climate science data and AI models to enhance infrastructure resilience planning. This involves predicting climate-related risks, optimizing designs for changing climate conditions, and developing adaptive infrastructure strategies.

3. **Transportation Engineering and Urban Planning:**

   - Application: Combining AI with transportation engineering principles to optimize traffic flow, design efficient transportation networks, and plan smart cities. AI is used for predictive modeling, traffic management, and the integration of different transportation modes.

4. **Geospatial Science and Infrastructure Planning:**

   - Application: Integrating geospatial data analytics with AI for improved infrastructure planning. This includes site selection, terrain analysis, and optimizing the placement of infrastructure components based on geographical and environmental considerations.

5. **Materials Science and Sustainable Construction:**

   - Application: Collaborating with materials science experts to develop AI-driven models for the discovery and optimization of sustainable construction materials. This interdisciplinary approach aims to enhance material performance while reducing environmental impact.

6. **Environmental Engineering and Smart Infrastructure:**

   - Application: Integrating environmental engineering principles with AI for the development of smart infrastructure. This involves real-time monitoring of environmental conditions, pollution detection, and adaptive control of infrastructure systems to minimize environmental impact.

7. **Human-Computer Interaction and Construction Robotics:**

   - Application: Applying principles from human-computer interaction to improve the collaboration between humans and construction robots. This includes developing user-friendly interfaces, intuitive controls, and human-robot interaction studies to enhance safety and efficiency.

8. **Economics and Project Management:**

   - Application: Incorporating economic principles into AI models for project management. This involves cost-benefit analysis, optimization of resource allocation, and risk assessment to improve the economic viability and financial sustainability of construction projects.

9. **Neuroscience and Human-Centric Design:**

   - Application: Leveraging insights from neuroscience to inform human-centric design in civil engineering projects. This interdisciplinary approach aims to create infrastructure that considers human behavior, well-being, and cognitive factors in its design and functionality.

10. **Data Science and Decision Support Systems:**

    - Application: Integrating advanced data science techniques with AI to develop decision support systems for civil engineering. This includes predictive analytics, data-driven risk assessment, and real-time monitoring to aid decision-making throughout the project lifecycle.

11. **Robotics and Structural Health Monitoring:**

    - Application: Combining robotics with structural health monitoring systems to create AI-driven solutions for inspecting and maintaining infrastructure. This interdisciplinary approach enhances the efficiency and accuracy of structural assessments.

12. **Medical Imaging and Non-Destructive Testing:**

    - Application: Adapting AI algorithms from medical imaging for non-destructive testing in civil engineering. This interdisciplinary approach enhances the detection of structural defects and assesses the integrity of materials without causing damage.

13. **Psychology and Safety Management:**

- Application: Integrating principles from psychology to enhance safety management in construction sites. AI models can be developed to predict and prevent unsafe conditions by understanding human behavior and psychological factors that contribute to accidents.

These interdisciplinary applications illustrate the diverse range of fields that can contribute to the advancement of AI in civil engineering. Collaborative efforts across disciplines can lead to more holistic and innovative solutions that address the multifaceted challenges faced by the civil engineering industry. As technology and research continue to progress, new interdisciplinary applications are likely to emerge, further expanding the possibilities for AI in civil engineering.

# 14.5 Speculations on Future Developments

While it's challenging to predict the future with certainty, we can speculate on potential developments in AI for civil engineering based on current trends and emerging technologies.

Here are some speculations on future developments in AI for civil engineering:

1. **AI-Integrated Construction Site Management:**
   - Speculation: AI will play a central role in real-time construction site management, coordinating tasks, optimizing workflows, and enhancing communication between on-site teams. Autonomous construction equipment and drones may

become commonplace, working collaboratively with human workers.

2. **AI-Driven Sustainable Design:**

- Speculation: AI algorithms will evolve to prioritize sustainability in the design phase of civil engineering projects. These algorithms will optimize designs for energy efficiency, material use, and environmental impact, contributing to the creation of more sustainable and eco-friendly infrastructure.

3. **AI-Enhanced Structural Health Monitoring:**

- Speculation: AI-powered structural health monitoring systems will become highly sophisticated, providing continuous and real-time assessments of the condition of infrastructure. Predictive maintenance models will enable proactive interventions to ensure the longevity and safety of structures.

4. **Generative Design Revolution:**

- Speculation: Generative design powered by AI will undergo a revolution, allowing civil engineers to explore an unprecedented number of design possibilities quickly. The emphasis will be on creating structures that are not only functional but aesthetically pleasing and optimized for various criteria, including cost and sustainability.

5. **AI for Rapid Urban Planning:**

- Speculation: AI will be instrumental in rapidly assessing and planning urban development projects. Advanced AI models will consider diverse factors such as population growth, environmental impact, traffic patterns, and social considerations to optimize urban infrastructure and enhance quality of life.

6. **Enhanced Human-Robot Collaboration:**

- Speculation: The collaboration between humans and robots on construction sites will become more seamless. AI algorithms will enable robots to adapt to dynamic environments, working alongside human workers to improve efficiency, safety, and the overall construction process.

7. **AI-Optimized Project Finance:**

   - Speculation: AI will be increasingly used in project finance for accurate cost estimations, risk assessments, and financial planning. AI algorithms will aid in optimizing budgets, reducing financial risks, and ensuring the economic viability of civil engineering projects.

8. **Cognitive Computing for Design Iterations:**

   - Speculation: Cognitive computing systems will facilitate rapid design iterations by understanding natural language input and interpreting design preferences. Engineers and architects will interact with AI systems conversationally, refining designs through an intuitive and collaborative process.

9. **AI-Driven Construction Materials Innovation:**

   - Speculation: AI will contribute to the discovery and development of innovative construction materials with enhanced properties. AI algorithms will analyze molecular structures, predict material behavior, and optimize compositions to meet specific performance criteria.

10. **Decentralized AI for Edge Computing:**

    - Speculation: Edge computing with decentralized AI will become more prevalent on construction sites, enabling real-time processing of data without relying heavily on centralized cloud infrastructure. This will lead to faster decision-making and improved responsiveness in construction operations.

11. **Quantum Computing for Simulation and Optimization:**

    - Speculation: As quantum computing technology matures, it may be applied to simulate complex structural behaviors and optimize large-scale civil engineering projects. Quantum algorithms could provide unprecedented computational capabilities for solving intricate optimization problems.

12. **Ethical AI Frameworks for Civil Engineering:**

    - Speculation: The industry will establish ethical frameworks specific to AI applications in civil engineering. These frameworks will address issues such as bias, transparency, and accountability, ensuring responsible and ethical use of AI technologies.

### 13. AI-Enabled Disaster Resilience Planning:

- Speculation: AI will contribute to advanced disaster resilience planning, helping cities and infrastructure adapt to and recover from natural disasters. Predictive models will assist in preparing for potential risks and developing resilient infrastructure strategies.

It's important to note that these speculations are based on current trends and potential technological advancements. The actual trajectory of AI in civil engineering will depend on various factors, including technological breakthroughs, regulatory developments, and societal acceptance. Continuous collaboration between researchers, industry professionals, and policymakers will play a crucial role in shaping the future landscape of AI in civil engineering.

\

# 14.6 Real-world Examples / Use Cases

1. **Future Innovations:**

   i. **AI-driven Generative Design (Company: Autodesk):**

      - Autodesk explores the future innovation of AI-driven generative design in civil engineering and construction. This involves using AI algorithms to generate multiple design options, optimizing for factors such as cost, sustainability, and functionality.

   ii. **Robotics and Autonomous Construction (Company: Boston Dynamics):**

      - Boston Dynamics envisions the future of AI in civil engineering with robotics and autonomous construction. Innovations include AI-powered construction robots capable of tasks such as excavation, bricklaying, and material handling.

2. **Emerging Technologies:**

   iii. **Digital Twin Integration (Company: Bentley Systems):**

      - Bentley Systems integrates digital twin technology with AI for emerging applications in construction. This involves creating real-time digital replicas of physical structures, enabling AI-driven simulations and predictive analytics.

   iv. **Blockchain for Construction (Company: Provenance):**

      - Provenance explores the use of blockchain technology with AI in construction. The integration ensures secure and transparent management of construction data, enhancing traceability and accountability in the industry.

3. **Research Frontiers:**

   v. **AI-driven Materials Innovation (Research: MIT Concrete Sustainability Hub):**

      - The MIT Concrete Sustainability Hub engages in research frontiers using AI to innovate construction materials. AI algorithms analyze material properties to develop eco-friendly and sustainable alternatives for construction.

vi. **AI-enhanced Safety Protocols (Research: CIFE at Stanford University):**

- The Center for Integrated Facility Engineering (CIFE) at Stanford University explores research frontiers in AI-enhanced safety protocols for construction. AI is applied to analyze on-site conditions, predict potential hazards, and improve overall safety measures.

4. **Industry Collaborations:**

vii. **Collaboration for AI-driven Project Planning (Collaboration: Turner Construction and Procore):**

- Turner Construction collaborates with Procore to implement AI-driven project planning. The partnership involves leveraging Procore's construction management platform with embedded AI features to enhance Turner's project planning processes.

viii. **AI Integration in Modular Construction (Collaboration: Katerra and Microsoft):**

- Katerra collaborates with Microsoft to integrate AI into modular construction processes. This collaboration aims to optimize design, manufacturing, and construction workflows using AI-driven insights.

5. **Interdisciplinary Applications:**

ix. **AI in Sustainable Urban Planning (Interdisciplinary: Sidewalk Labs):**

- Sidewalk Labs explores interdisciplinary applications of AI in civil engineering for sustainable urban planning. The project involves using AI to analyze data related to transportation, infrastructure, and energy usage to optimize urban design.

x. **AI for Infrastructure Resilience (Interdisciplinary: Resilience Shift):**

- The Resilience Shift initiative explores interdisciplinary applications of AI for infrastructure resilience. By integrating AI algorithms, the project aims to enhance the resilience of critical infrastructure systems to withstand various challenges.

## 6. Speculations on Future Developments:

### xi. AI for Self-Healing Structures:

- Future developments in AI for civil engineering may include self-healing structures. AI algorithms could be employed to monitor and repair structural damage autonomously, ensuring the longevity and sustainability of infrastructure.

### xii. AI-driven Augmented Reality Construction Assistants:

- Speculations involve the development of AI-driven augmented reality construction assistants. These assistants would provide on-site workers with real-time information, guidance, and data overlays, enhancing efficiency and reducing errors in construction tasks.

## 14.7 Chapter Summary: Key Points

1. AI's role in civil engineering is marked by advancements in Generative Design, where algorithms optimize design alternatives, fostering creativity and efficiency.

2. Predictive maintenance using AI is evolving, leveraging sophisticated machine learning models for equipment failure prediction, optimizing schedules, and minimizing downtime.

3. Autonomous construction machinery, integrated with AI, promises increased efficiency and safety, allowing real-time decision-making and collaboration with human workers.

4. AI-driven generative design tools aid architects and engineers in creating optimized, sustainable building designs prioritizing energy efficiency.

5. Enhanced predictive analytics using AI algorithms revolutionizes project planning, providing accurate forecasts of timelines, costs, and potential risks.

6. Smart construction materials with self-monitoring capabilities, adaptive properties, and self-repairing features are anticipated innovations in civil engineering.

7. Drones and AI integration for real-time construction site monitoring, AI-driven structural health monitoring, and digital twin technology integration offer transformative advancements.

8. AI in urban planning optimizes city layouts, transportation systems, and infrastructure development for enhanced efficiency and sustainability.

9. Natural language processing integration into project communication systems and AI-driven HVAC systems for energy-efficient building operation are future trends.

10. AI's role in autonomous construction machinery promises increased efficiency, safety, and collaboration between machines and human workers.

11. AI integration with Building Information Modeling (BIM) transforms project planning and management by analyzing vast datasets for insights.

12. Real-time collaboration in construction projects advances with AI-driven tools facilitating communication and coordination among stakeholders.

13. AI-driven safety monitoring systems analyze wearable devices and video feeds to detect safety hazards, predict risky behaviors, and prevent accidents.

14. AI in materials science leads to innovations in advanced construction materials, optimizing durability, sustainability, and structural integrity.

15. AI applications in risk assessment, environmental sustainability, and interdisciplinary collaborations illustrate the transformative impact of AI on civil engineering practices.

## 14.8 Concept Check: Q&A Sessions

1. What is Generative Design, and how does it impact civil engineering?

2. How does AI contribute to predictive maintenance in civil engineering?

3. In what ways is AI integrated into construction machinery, and how does it impact efficiency and safety?

4. What role does AI play in project planning, and how does it revolutionize forecasting in civil engineering?

5. How are smart construction materials expected to evolve with the integration of AI?

6. What are the key advancements in the combined use of drones and AI in construction site monitoring?

7. How does AI contribute to structural health monitoring, and what advancements are expected?

8. In what ways is AI expected to influence urban planning, and what aspects of city development does it optimize?

9. How is natural language processing (NLP) integrated into project communication systems in civil engineering?

10. What role does AI play in the development of HVAC systems for energy-efficient building operation?

# 15. Ethical Considerations in AI for Construction

Ethical considerations in the use of Artificial Intelligence (AI) for construction projects are of paramount importance as these technologies become integral to the industry.

One primary ethical concern is the potential bias in AI algorithms. If these algorithms are trained on biased datasets, they may perpetuate and amplify existing social and economic inequalities. Therefore, ensuring diverse and representative data for training AI models is crucial to avoid discriminatory outcomes.

Transparency in AI decision-making processes is another ethical consideration. Construction stakeholders must understand how AI systems arrive at decisions, especially when those decisions have significant implications for project outcomes. Clear communication about the role of AI, the data used, and the decision-making criteria helps in building trust among project participants.

Data privacy is a critical ethical concern in AI for construction. As AI systems process vast amounts of data, ensuring that sensitive information about individuals, companies, or projects is appropriately protected is imperative. Robust data governance practices, including encryption and secure storage, are essential to uphold privacy standards.

Responsible AI use in construction demands a commitment to safety. Autonomous construction machinery and robotic systems driven by AI should adhere to strict safety standards to avoid accidents or injuries. Ethical considerations include developing fail-safe mechanisms and implementing rigorous testing procedures to ensure the safety of both workers and the public.

Equitable access to AI technologies is an ethical consideration that must be addressed. Small and medium-sized construction firms should not be disproportionately disadvantaged due to limited access to AI resources. Ensuring accessibility and affordability of AI tools promotes fair competition and a level playing field within the industry.

AI in construction must be aligned with legal and regulatory frameworks. Ethical considerations include compliance with local and international laws, industry standards, and codes of conduct. Adhering to these regulations helps prevent legal challenges, promotes ethical behavior, and ensures responsible AI deployment.

The displacement of jobs due to increased automation in construction is an ethical concern that demands attention. While AI technologies can improve efficiency, it is crucial to address potential job losses and invest in reskilling and upskilling programs to transition workers into roles that complement AI systems.

Environmental sustainability is an ethical imperative in construction AI. Ensuring that AI-driven decisions consider environmental impact, resource conservation, and sustainable practices aligns with ethical considerations for responsible and eco-friendly construction practices.

Ongoing monitoring and auditing of AI systems are ethical responsibilities. Regular assessments of AI models, algorithms, and their impact on construction projects are essential to identify and rectify any unintended consequences or biases that may emerge over time.

Promoting a culture of ethical awareness and accountability within construction organizations is vital. Ensuring that all stakeholders, from engineers to project managers, understand the ethical implications of AI adoption fosters a responsible and ethical approach to AI use in construction. Ethical considerations should be integrated into training programs and corporate policies to guide ethical decision-making throughout the AI lifecycle in construction projects.

## 15.1 Bias and Fairness

Addressing bias and ensuring fairness in AI for construction is crucial for ethical considerations.

Bias in AI models can lead to discriminatory outcomes, perpetuate inequalities, and undermine the trustworthiness of AI applications. Here are key considerations and strategies to mitigate bias and enhance fairness in AI for construction:

1.  **Data Quality and Bias Detection:**

    - Consideration: Biases often stem from biased training data. Ensure that training datasets used to develop AI models are diverse, representative, and free from biases.

- Strategy: Implement thorough data quality assessments and conduct bias detection analyses regularly. Address and correct biases in the training data to promote fairness in AI outcomes.

2. **Diverse Representation:**

   - Consideration: Ensure that the development and validation datasets include diverse representation of construction scenarios, environments, and demographics.

   - Strategy: Actively seek input from diverse stakeholders in the construction industry to contribute to the development and testing of AI models. This includes input from construction workers, engineers, project managers, and other relevant professionals.

3. **Explainability and Transparency:**

   - Consideration: Lack of transparency in AI decision-making can lead to distrust. Understandability of AI decisions is essential for accountability.

   - Strategy: Use interpretable models and provide explanations for AI predictions and decisions. Make AI processes and algorithms transparent to end-users and stakeholders, fostering trust and accountability.

4. **Fairness Metrics and Audits:**

   - Consideration: Define fairness metrics that align with ethical standards and industry norms. Regularly audit AI models for fairness to identify and rectify potential biases.

   - Strategy: Employ fairness metrics, such as disparate impact analysis, to assess and measure biases in AI outputs. Conduct regular audits to ensure ongoing fairness in AI applications.

5. **Inclusive Design and User Feedback:**

   - Consideration: Inclusive design involves considering the needs of all potential users. User feedback is essential for understanding the impact of AI on diverse users.

   - Strategy: Involve diverse stakeholders in the design process, gather user feedback, and iterate on AI models based on the input received. Prioritize inclusivity to avoid unintentional biases.

6. **Bias Mitigation Techniques:**

- Consideration: Employ techniques to actively mitigate biases during the development and deployment of AI models.

- Strategy: Explore methods such as re-sampling, re-weighting, and adversarial training to reduce biases. Regularly update models to incorporate evolving best practices in bias mitigation.

7. **Ethical Guidelines and Policies:**

- Consideration: Establish clear ethical guidelines and policies for the development and use of AI in construction.

- Strategy: Develop and communicate ethical guidelines that explicitly address bias, fairness, and the responsible use of AI in construction. Enforce these guidelines through policies that prioritize ethical considerations.

8. **Diversity in Development Teams:**

- Consideration: The composition of development teams can influence the perspectives embedded in AI models.

- Strategy: Promote diversity within AI development teams to bring in a variety of perspectives and experiences. Diverse teams are more likely to identify and address potential biases during the development process.

9. **Continuous Monitoring and Updating:**

- Consideration: AI models can become biased over time as construction practices, technologies, and demographics evolve.

- Strategy: Implement continuous monitoring of AI models in real-world conditions. Regularly update models to adapt to changes in the construction industry and ensure ongoing fairness.

10. **External Audits and Certification:**

- Consideration: External validation can provide an independent assessment of AI fairness.

- Strategy: Engage third-party auditors or certification bodies to assess the fairness of AI models used in construction. External scrutiny can provide additional assurance and accountability.

By integrating these considerations and strategies into the development and deployment of AI for construction, stakeholders can work towards mitigating bias, ensuring fairness, and promoting the responsible and ethical use of AI technologies in the industry. Regular engagement with the construction community, along with a commitment to transparency, is essential for building trust in AI applications.

## 15.2 Transparency

Transparency is a critical aspect of ethical considerations in AI for construction.

It involves making AI models, algorithms, and decision-making processes understandable and clear to users, stakeholders, and the broader public. Ensuring transparency in AI for construction contributes to accountability, trust-building, and ethical use. Here are key considerations and strategies for promoting transparency in AI for construction:

1. **Explainability of Models:**

   o Consideration: Users and stakeholders should be able to understand how AI models make decisions and predictions.

o Strategy: Choose models that offer explainability, and provide clear explanations for AI outputs. Techniques such as local interpretable model-agnostic explanations (LIME) and SHapley Additive exPlanations (SHAP) can be employed to interpret model predictions.

2. **Documentation and Model Metadata:**

o Consideration: Comprehensive documentation enhances understanding of AI models and their deployment.

o Strategy: Document the development process, including data sources, preprocessing steps, and model architecture. Provide metadata that details the model's purpose, limitations, and potential biases. This documentation should be easily accessible to relevant stakeholders.

3. **User-Friendly Interfaces:**

o Consideration: Users interacting with AI systems, such as construction professionals, should have intuitive and user-friendly interfaces.

o Strategy: Design interfaces that present AI outputs in an understandable manner. Use visualizations and clear language to communicate information, making it accessible to users without a technical background.

4. **Transparency in Decision-Making:**

o Consideration: Transparency should extend to the decision-making processes influenced by AI in construction projects.

o Strategy: Clearly communicate how AI recommendations or decisions integrate with overall project decision-making. Outline the factors considered by AI models and their role in informing decisions.

5. **Open Source and Collaboration:**

o Consideration: Open-source projects enable transparency by allowing external scrutiny and collaboration.

o Strategy: Consider open-sourcing non-proprietary components of AI systems used in construction. Collaboration with the wider AI and construction communities fosters transparency through shared knowledge and expertise.

6. **Data Provenance and Traceability:**

   o Consideration: Knowing the origin and history of data is crucial for understanding AI outputs.

   o Strategy: Establish data provenance practices that trace the origin, transformations, and usage of data throughout the AI development process. This enhances accountability and transparency in case of issues.

7. **Regular Reporting and Audits:**

   o Consideration: Regular reporting and audits provide ongoing transparency into AI model performance.

   o Strategy: Establish a schedule for reporting on AI model performance, accuracy, and any updates or changes. Conduct periodic audits, both internally and externally, to ensure ongoing transparency and adherence to ethical standards.

8. **Ethical Guidelines and Policies:**

   o Consideration: Clearly defined ethical guidelines and policies contribute to transparency.

   o Strategy: Develop and communicate transparent ethical guidelines for AI use in construction. These guidelines should address data handling, model deployment, and considerations for ethical decision-making.

9. **Engagement with Stakeholders:**

   o Consideration: Engaging with stakeholders fosters transparency by providing them insights into the AI systems.

   o Strategy: Involve construction professionals, project managers, and other stakeholders in the AI development process. Seek and incorporate their input to enhance transparency and align AI applications with real-world needs.

10. **Education and Training:**

   o Consideration: Users need to be educated on how to interpret and interact with AI systems.

- o Strategy: Provide training and educational resources to users to enhance their understanding of AI models, their capabilities, and their limitations. Promote a culture of continuous learning and transparency.

## 11. Communication of Updates:

- o Consideration: Regularly updating stakeholders on changes in AI models, algorithms, or policies is essential.

- o Strategy: Communicate transparently about updates, improvements, and potential changes in AI systems. Be proactive in addressing concerns and clarifying any modifications to the AI infrastructure.

## 12. Compliance with Regulations:

- o Consideration: Compliance with data protection and privacy regulations enhances transparency and legal accountability.

- o Strategy: Ensure that AI applications comply with relevant regulations, such as GDPR, and communicate how data privacy and security are prioritized in the design and deployment of AI systems.

By incorporating these considerations and strategies, construction professionals and AI practitioners can work towards building transparency into AI systems used in the construction industry. This not only aligns with ethical principles but also contributes to the responsible and trustworthy deployment of AI technologies in construction projects.

## 15.3 Accountability

Ensuring accountability is a key component of ethical considerations in AI for construction.

Accountability involves establishing clear lines of responsibility, transparency in decision-making processes, and mechanisms to address the consequences of AI applications. Here are essential considerations and strategies for promoting accountability in AI for construction:

1.  **Clearly Defined Roles and Responsibilities:**
    o   Consideration: Clearly define the roles and responsibilities of individuals involved in the development, deployment, and oversight of AI systems in construction.

o Strategy: Establish accountability frameworks that outline the specific responsibilities of data scientists, engineers, project managers, and other stakeholders. Clearly communicate these roles to ensure accountability at each stage of AI development and use.

2. **Ethical Guidelines and Codes of Conduct:**

o Consideration: Develop and communicate ethical guidelines and codes of conduct that align with industry standards and ethical principles.

o Strategy: Define a set of ethical guidelines specific to AI applications in construction. These guidelines should encompass data privacy, fairness, transparency, and other ethical considerations. Ensure that all stakeholders are aware of and adhere to these guidelines.

3. **Oversight and Governance Structures:**

o Consideration: Establish oversight and governance structures to monitor the development and deployment of AI systems.

o Strategy: Implement governance mechanisms that include regular reviews, audits, and oversight by designated teams or individuals. These structures ensure that AI applications align with ethical standards and industry regulations.

4. **Compliance with Regulations:**

o Consideration: Ensure compliance with relevant laws, regulations, and industry standards related to AI and data privacy.

o Strategy: Stay informed about existing and emerging regulations governing AI in construction. Implement processes and controls to ensure compliance with legal and regulatory requirements.

5. **Documentation and Accountability Reports:**

o Consideration: Maintain comprehensive documentation of AI development processes and outcomes.

o Strategy: Document the entire AI lifecycle, including data sources, model development, testing, and deployment. Generate accountability reports that highlight key decisions, model performance, and adherence to ethical guidelines. These reports contribute to transparency and accountability.

6. **Training and Awareness Programs:**

o Consideration: Ensure that individuals involved in AI development and use are adequately trained and aware of ethical considerations.

o Strategy: Conduct regular training programs to educate team members on ethical principles, responsible AI practices, and potential risks. Build awareness of the importance of accountability throughout the organization.

7. **User Feedback and Input:**

o Consideration: Incorporate user feedback and input to hold AI systems accountable for meeting user needs and expectations.

o Strategy: Establish channels for collecting feedback from construction professionals, workers, and other stakeholders. Actively seek input to identify areas for improvement and to ensure that AI applications align with user requirements.

8. **Continuous Monitoring and Evaluation:**

o Consideration: Implement continuous monitoring and evaluation mechanisms to assess the ongoing performance of AI systems.

o Strategy: Regularly evaluate the accuracy, fairness, and ethical implications of AI models. Use real-world feedback and performance metrics to identify and address issues promptly.

9. **Incident Response and Redress Mechanisms:**

o Consideration: Prepare for and respond to incidents related to AI system failures, biases, or ethical breaches.

o Strategy: Establish incident response plans and redress mechanisms to address any unintended consequences or ethical lapses. Communicate transparently about incidents and take corrective actions promptly.

## 10. External Audits and Certification:

- o Consideration: External validation through audits and certifications enhances accountability.

- o Strategy: Engage external auditors or certification bodies to assess the adherence of AI systems to ethical guidelines and industry standards. External scrutiny provides an additional layer of accountability.

## 11. Legal and Contractual Accountability:

- o Consideration: Clearly outline legal and contractual responsibilities related to AI applications in construction.

- o Strategy: Ensure that contracts, agreements, and service-level agreements explicitly define the legal responsibilities of each party involved in the development and deployment of AI systems. This includes liability considerations and dispute resolution mechanisms.

## 12. Public Accountability and Communication:

- o Consideration: Foster public accountability by communicating openly with the public about AI systems and their impact.

- o Strategy: Engage in transparent communication about the purpose, benefits, and potential risks of AI applications in construction. Address public concerns and maintain accountability through open dialogue.

By implementing these considerations and strategies, stakeholders in the construction industry can establish a robust framework for accountability in AI development and deployment. This framework ensures that ethical principles are upheld, and responsible practices are followed throughout the lifecycle of AI applications in construction projects.

## 15.4 Privacy Concerns

Addressing privacy concerns is a critical aspect of ethical considerations in AI for construction, especially given the sensitivity of the data involved in construction projects.

Privacy issues can arise from the collection, storage, and processing of personal or sensitive information. Here are key considerations and strategies to mitigate privacy concerns in AI for construction:

1.  **Data Minimization:**
    o   Consideration: Minimize the collection and use of personally identifiable information (PII) to only what is necessary for the intended purpose.

o Strategy: Adopt a data minimization approach by collecting only the data required for specific tasks. Avoid unnecessary or excessive data collection to reduce the potential impact on privacy.

2. **Anonymization and Pseudonymization:**

o Consideration: Protect the privacy of individuals by anonymizing or pseudonymizing personal data.

o Strategy: Implement techniques such as anonymization (removing personally identifiable information) or pseudonymization (replacing identifiable information with pseudonyms) to reduce the risk of identifying individuals from the data.

3. **Secure Data Storage and Transmission:**

o Consideration: Ensure the security of data storage and transmission to prevent unauthorized access and data breaches.

o Strategy: Use secure and encrypted storage systems. Implement encryption protocols for data transmission to protect sensitive information from interception or unauthorized access.

4. **Informed Consent:**

o Consideration: Obtain informed consent from individuals before collecting and using their personal data.

o Strategy: Clearly communicate the purpose, scope, and potential uses of data to individuals. Seek explicit consent, and provide mechanisms for individuals to opt-in or opt-out of data collection and processing activities.

5. **User Control and Access Rights:**

o Consideration: Empower individuals with control over their data and provide access rights to manage their personal information.

o Strategy: Develop user interfaces that allow individuals to review, edit, or delete their data. Clearly communicate how individuals can exercise their rights concerning their personal information.

6. **Privacy Impact Assessments (PIA):**

   o Consideration: Conduct privacy impact assessments to identify and address potential privacy risks associated with AI applications.

   o Strategy: Integrate privacy impact assessments into the development process. Assess the impact of AI systems on privacy, and implement measures to mitigate identified risks.

7. **Ethical Data Handling Practices:**

   o Consideration: Adhere to ethical data handling practices that prioritize privacy and respect for individuals.

   o Strategy: Establish clear guidelines for ethical data handling, including data access controls, sharing protocols, and retention policies. Ensure that all personnel involved in AI projects are trained on ethical data practices.

8. **Data Ownership and Accountability:**

   o Consideration: Clarify data ownership and establish accountability for the handling of personal data.

   o Strategy: Clearly define data ownership rights and responsibilities in contracts and agreements. Assign accountability to specific individuals or teams for ensuring compliance with privacy regulations and ethical standards.

9. **Regulatory Compliance:**

   o Consideration: Stay compliant with relevant data protection and privacy regulations, such as GDPR.

   o Strategy: Regularly monitor changes in privacy regulations and ensure that AI applications comply with legal requirements. Appoint a data protection officer if necessary and establish processes for reporting and managing privacy breaches.

10. **Transparency and Communication:**

   o Consideration: Communicate transparently with individuals about how their data is being used and processed.

   o Strategy: Provide clear and understandable privacy policies. Communicate openly about data practices, including the role of

AI, and address privacy concerns raised by individuals or the public.

## 11. Regular Audits and Assessments:

- o Consideration: Conduct regular audits and assessments to evaluate the effectiveness of privacy safeguards.

- o Strategy: Perform internal and external audits to assess compliance with privacy policies and regulations. Use the findings to continuously improve privacy measures and address any identified weaknesses.

## 12. Educating Stakeholders:

- o Consideration: Educate all stakeholders, including employees, contractors, and users, about the importance of privacy in AI applications.

- o Strategy: Provide training programs on privacy considerations and data protection practices. Foster a culture of privacy awareness within the organization.

By incorporating these considerations and strategies, construction industry stakeholders can work towards mitigating privacy concerns associated with AI applications. Prioritizing privacy not only aligns with ethical principles but also contributes to building trust with individuals whose data is involved in construction-related AI projects.

## 15.5 Social Impacts of AI in Construction

The integration of Artificial Intelligence (AI) in the construction industry brings about various social impacts, influencing how projects are planned, executed, and managed.

Here are some notable social impacts of AI in construction:

1. **Improved Safety:**

   o Impact: AI applications enhance safety by predicting and preventing potential hazards on construction sites.

   o Explanation: AI-enabled sensors and cameras can monitor work environments in real-time, identifying safety risks and alerting workers to potential dangers. Predictive analytics can also help in planning safer construction processes.

2. **Enhanced Productivity and Efficiency:**

   o Impact: AI technologies contribute to increased efficiency, leading to faster project completion.

   o Explanation: Automation of repetitive tasks, optimized scheduling, and resource management through AI-driven algorithms result in improved productivity. This can lead to quicker project timelines and potentially reduce labor-intensive work.

3. **Skill Requirements and Workforce Upskilling:**

   o Impact: The adoption of AI in construction may require upskilling the workforce to handle new technologies.

   o Explanation: Workers may need to learn how to operate and collaborate with AI-driven systems. This creates opportunities for ongoing training and skill development, contributing to a more knowledgeable and adaptable workforce.

4. **Job Transformation and Creation:**

   o Impact: AI in construction can transform job roles, creating new positions while modifying existing ones.

   o Explanation: While some routine tasks may become automated, the demand for AI system operators, data analysts, and maintenance personnel may increase. This shift can lead to a redistribution of job responsibilities and the creation of new roles.

5. **Data-Driven Decision-Making:**

   o Impact: AI facilitates data-driven decision-making processes in construction projects.

   o Explanation: Access to real-time data and analytics helps project managers and stakeholders make informed decisions. This can lead to more accurate resource allocation, better risk management, and improved project outcomes.

6. **Remote Collaboration and Communication:**

- o Impact: AI supports remote collaboration and communication, reducing the need for physical presence on construction sites.

- o Explanation: AI-powered communication tools, virtual reality (VR), and augmented reality (AR) enable remote collaboration among teams. This can lead to increased flexibility in work arrangements and a more geographically dispersed workforce.

7. **Affordability and Accessibility:**

- o Impact: AI technologies become more affordable and accessible, democratizing their use in construction projects.

- o Explanation: As AI solutions become more commonplace and costs decrease, smaller construction firms and projects may also benefit from AI-driven tools, fostering more widespread adoption across the industry.

8. **Environmental Sustainability:**

- o Impact: AI contributes to more sustainable construction practices.

- o Explanation: AI can optimize energy usage, reduce waste, and enhance the design of eco-friendly structures. This aligns with societal concerns for environmentally responsible construction practices.

9. **Community Engagement and Social Responsibility:**

- o Impact: AI aids in better community engagement and social responsibility initiatives.

- o Explanation: AI technologies can be utilized to collect and analyze community feedback, fostering a more inclusive decision-making process. Construction companies may use AI to assess and address the social impact of their projects on local communities.

10. **Ethical Considerations and Inclusivity:**

- o Impact: The ethical use of AI in construction becomes a priority, focusing on inclusivity and fairness.

- o Explanation: Construction companies adopting AI are expected to address ethical considerations, ensuring that AI systems do not contribute to biases or exclusion. Fair hiring practices and inclusive AI algorithms become crucial for social impact.

### 11. Public Perception and Acceptance:

○ Impact: Public perception of AI in construction influences its acceptance and integration.

○ Explanation: The construction industry needs to communicate the benefits and safeguards of AI adoption to the public. Establishing trust and transparency regarding AI applications can positively impact societal acceptance.

### 12. Urban Planning and Smart Cities:

○ Impact: AI influences urban planning and the development of smart cities.

○ Explanation: AI technologies play a role in designing and managing infrastructure for smart cities, addressing societal needs such as efficient transportation, energy usage, and overall urban sustainability.

It's important for stakeholders in the construction industry to actively consider and address the social impacts of AI. This involves implementing ethical guidelines, fostering collaboration, and ensuring that the adoption of AI technologies aligns with societal values and concerns. Continuous dialogue between industry players, regulators, and the public is crucial to navigate these social impacts responsibly. Subsequent chapters will discuss various cases studies with reference to usage of AI in Constructions.

## 15.6 Real-world Examples / Use Cases

1. **Ethical Considerations in AI for Civil Engineering and Construction:**

   **Turner Construction Company** has implemented ethical guidelines that prioritize the safety and well-being of workers when deploying AI-driven construction machinery, ensuring responsible and ethical use in their projects.

2. **Bias and Fairness in AI for Civil Engineering and Construction:**

   Skanska, a global construction company, actively addresses bias by incorporating diverse perspectives from construction professionals, engineers, and project managers in the development and testing of AI models, aiming for fair and unbiased outcomes.

3. **Transparency in AI for Civil Engineering and Construction:**

   Autodesk, through its AI-powered construction planning tools, ensures transparency by providing clear explanations for AI-generated project schedules and decisions, fostering trust among construction stakeholders through understandable and transparent AI processes.

4. **Accountability in AI for Civil Engineering and Construction:**

   AECOM, a multinational engineering firm, establishes accountability frameworks that clearly define the roles and responsibilities of data scientists, engineers, and project managers in the development and oversight of AI systems, ensuring accountability at each stage of AI implementation.

5. **Privacy Concerns in AI for Civil Engineering and Construction:**

   Bentley Systems, a software development company, addresses privacy concerns by implementing stringent data minimization practices, collecting only necessary data for their AI-driven infrastructure management systems and prioritizing user consent for data processing.

6. **Social Impacts of AI in Civil Engineering and Construction:**

China State Construction Engineering Corporation (CSCEC) integrates AI for safety monitoring on construction sites, positively impacting the workforce by improving safety conditions and reducing accidents, demonstrating the social benefits of AI in construction.

## 15.7 Chapter Summary: Key Points

1. Ethical AI in construction requires diverse and representative data to avoid perpetuating existing inequalities.

2. Clear communication about AI roles, data use, and decision criteria is essential for building trust among construction stakeholders.

3. Robust data governance practices, including encryption and secure storage, are imperative to protect sensitive information in AI for construction.

4. Autonomous construction machinery driven by AI must adhere to strict safety standards to prevent accidents and injuries.

5. Ensuring small and medium-sized construction firms have access to and can afford AI resources promotes fair competition in the industry.

6. Adherence to local and international laws, industry standards, and codes of conduct is crucial to prevent legal challenges in AI for construction.

7. Addressing potential job losses through increased automation requires investment in reskilling and upskilling programs for workers.

8. Ethical AI in construction involves considering environmental impact, resource conservation, and sustainable practices.

9. Regular assessments of AI models and algorithms are essential to identify and rectify unintended consequences or biases.

10. Promoting a culture of ethical awareness within construction organizations is vital for responsible AI use.

11. Strategies include diverse representation, explainability, fairness metrics, inclusive design, and continuous monitoring to address bias in construction AI.

12. Explainability, documentation, user-friendly interfaces, and open-source collaboration contribute to transparency in construction AI.

13. Clearly defined roles, ethical guidelines, oversight structures, and compliance mechanisms establish accountability in AI for construction.

14. Data minimization, anonymization, secure storage, informed consent, and regular audits mitigate privacy concerns in construction AI.

15. Improved safety, enhanced productivity, upskilling, job transformation, data-driven decision-making, remote collaboration, affordability, sustainability, community engagement, ethical considerations, public perception, and smart city development are key social impacts of AI in construction.

## 15.8 Concept Check: Q&A Sessions

1. Why is addressing bias and ensuring fairness crucial in AI for construction?

2. How can biases in AI models be mitigated in the construction industry?

3. Why is diversity in development teams considered important in addressing bias in AI for construction?

4. How does transparency contribute to ethical considerations in AI for construction?

5. What role do explainable models play in promoting transparency in AI for construction?

6. How can AI practitioners ensure transparency in decision-making processes related to construction projects?

7. What is the significance of open source and collaboration in promoting transparency in AI for construction?

8. How does data provenance contribute to transparency in AI for construction?

9. Why is continuous monitoring and updating of AI models important in ensuring ongoing fairness?

10. How can construction professionals address privacy concerns associated with AI applications?

# 16. AI driven Construction Projects and Case Studies

Artificial Intelligence (AI) for construction requires a strategic and systematic approach to harness the full potential of these technologies.

Before we discuss various case studies in the subsequent chapters, we shall explore the system approaches in AI driven construction projects. Firstly, a thorough needs assessment should be conducted to identify specific pain points and opportunities for improvement within construction processes. This involves understanding existing workflows, project complexities, and the unique challenges faced by construction teams.

Following the needs assessment, a clear and comprehensive AI strategy should be formulated. This strategy should outline specific goals, expected outcomes, and a timeline for implementation. Prioritizing AI applications based on their potential impact on efficiency, cost savings, and overall project success is crucial in developing a focused and effective strategy.

Collaboration and communication are key elements in the successful implementation of AI for construction. Stakeholders, including project managers, engineers, and field workers, should be actively involved in the process. Open communication channels facilitate a shared understanding of AI's role, benefits, and potential challenges, fostering a collaborative approach to implementation.

Investing in talent and technology is a fundamental aspect of AI implementation. Construction firms need to identify the right skill sets for managing AI initiatives and provide training for their workforce. Additionally, selecting appropriate AI technologies and tools that align with the firm's specific needs is crucial for successful implementation.

Piloting AI initiatives on a smaller scale allows construction firms to test and validate the technology before broader adoption. AI driven pilot construction projects helps in identifying challenges, fine-tuning processes, and building confidence among stakeholders regarding the benefits of AI in construction workflows.

Integration with existing technologies and systems is a key consideration in AI implementation. Construction firms should ensure that AI technologies seamlessly interface with their current software, hardware, and data infrastructure. Compatibility and interoperability are essential for a smooth transition and effective utilization of AI capabilities.

Data management is a critical aspect of AI implementation in construction. Establishing robust data governance practices ensures data quality, security, and accessibility. A well-organized data infrastructure forms the foundation for AI applications, allowing algorithms to operate on reliable and relevant information.

Building partnerships with AI solution providers and technology experts can accelerate the implementation process. Collaboration with industry specialists facilitates knowledge transfer, access to cutting-edge technologies, and ongoing support for overcoming challenges encountered during implementation.

## 16.1 Large-scale Infrastructure Projects

AI in large-scale infrastructure projects demonstrate how AI-driven technologies can enhance efficiency, safety, and overall project management in the construction industry.

Here are some notable examples:

1. **Heathrow Airport Expansion (London, UK):**

   o **AI Application:** AI is used for project management, scheduling, and resource optimization.

   The expansion project at Heathrow Airport leverages AI to optimize construction schedules, predict potential delays, and allocate resources efficiently. This improves overall project timelines and resource utilization.

2. **Hong Kong-Zhuhai-Macau Bridge (Hong Kong, China):**

   o **AI Application:** AI is employed for structural health monitoring and predictive maintenance.

   AI sensors are used to monitor the structural health of the bridge in real-time. Predictive maintenance algorithms analyze data to identify potential issues before they become critical, ensuring the safety and integrity of the structure.

3. **Grand Paris Express Metro (Paris, France):**

   o **AI Application:** AI is utilized for tunneling optimization and geospatial analysis.

   AI algorithms optimize the tunneling process by analyzing geological data in real-time. This improves the efficiency of tunnel construction and minimizes the risk of delays due to unexpected geological conditions.

4. **California High-Speed Rail (California, USA):**

   o **AI Application:** AI is used for project management, risk assessment, and route optimization.

   AI tools assist in project planning, risk analysis, and route optimization for the high-speed rail project in California. This enhances decision-making, reduces project risks, and improves overall project efficiency.

5. **Singapore Changi Airport Terminal 5 (Singapore):**

   o **AI Application:** AI is employed for design optimization and energy efficiency.

   AI-driven design optimization tools are used to enhance the architectural and engineering aspects of the terminal. Additionally, AI algorithms optimize energy consumption, contributing to the sustainability of the infrastructure.

6. **Doha Metro (Doha, Qatar):**

   o **AI Application:** AI is utilized for traffic management and passenger flow analysis.

AI systems monitor and analyze passenger flows in real-time, optimizing train schedules and station operations. This ensures efficient transportation services and a seamless passenger experience.

7. **Crossrail Project (London, UK):**

   o **AI Application:** AI is employed for predictive maintenance and safety monitoring.

   AI sensors and predictive maintenance algorithms are used to monitor the condition of tracks, tunnels, and other infrastructure elements. This proactive approach to maintenance improves safety and minimizes disruptions.

8. **Oslo Airport Expansion (Oslo, Norway):**

   o **AI Application:** AI is used for construction site safety and security.

   AI-based video analytics are employed to monitor construction site safety and security. The system can identify potential safety hazards and security breaches, contributing to a safer working environment.

9. **High-Speed 2 (HS2) Rail Project (United Kingdom):**

   o **AI Application:** AI is applied for project planning and environmental impact assessment.

   AI tools are used to analyze project plans, assess potential environmental impacts, and optimize routes to minimize ecological disruptions. This supports sustainable and environmentally conscious construction practices.

These case studies highlight how AI is being applied across different aspects of large-scale infrastructure projects, including project management, safety monitoring, design optimization, and environmental impact assessment. As the field of AI in construction continues to evolve, it is likely that more projects will leverage these technologies to improve efficiency, sustainability, and safety in construction processes.

## 16.2 Residential and Commercial Developments

There are several case studies showcase the application of AI in residential and commercial developments.

AI-driven construction projects in these sectors often focus on optimizing processes, improving project management, enhancing energy efficiency, and ensuring sustainability. Here are some notable case studies:

1. **Sidewalk Labs - Quayside (Toronto, Canada):**

   o **AI Application:** AI is used for smart city planning and infrastructure development.

   Sidewalk Labs, an Alphabet subsidiary, is developing Quayside as a smart city project. AI technologies are applied to optimize urban planning, traffic flow, and environmental sustainability, creating a technologically advanced and efficient urban environment.

2. **Smart Dubai - Al Wasl Plaza (Dubai, UAE):**

   o **AI Application:** AI is utilized for smart building management and energy efficiency.

   Al Wasl Plaza, a key component of the Expo 2020 Dubai site, incorporates AI for smart building systems. The technology is used to optimize energy consumption, improve operational efficiency, and enhance the overall sustainability of the development.

3. **One Central Park (Sydney, Australia):**

   o **AI Application:** AI is applied for energy-efficient design and green building features.

   One Central Park, a residential and commercial development, integrates AI for designing energy-efficient buildings. Automated shading systems, powered by AI algorithms, adjust based on sunlight conditions to optimize natural lighting and reduce energy consumption.

4. **Shanghai Tower (Shanghai, China):**

   o **AI Application:** AI is used for building management, predictive maintenance, and energy optimization.

   Shanghai Tower, one of the world's tallest buildings, utilizes AI for various aspects of building management. Predictive maintenance algorithms are employed to anticipate equipment failures, and energy optimization strategies contribute to sustainability.

5. **175 West 60th Street (New York, USA):**

   o **AI Application:** AI is applied for construction site safety and monitoring.

   AI-based monitoring systems are implemented to enhance construction site safety. Video analytics and sensors help identify potential safety hazards in real-time, allowing for proactive interventions to ensure a secure working environment.

6. **The Edge (Amsterdam, Netherlands):**

   o **AI Application:** AI is utilized for smart building operations and user experience.

   The Edge, an office building, employs AI for smart lighting, heating, and other building systems. The technology adapts to user preferences, optimizing the working environment and contributing to energy efficiency.

7. **Optimizing Modular Construction (Various Locations):**

   o **AI Application:** AI is used to optimize modular construction processes.

   In various residential and commercial projects, AI is applied to optimize modular construction workflows. Machine learning algorithms analyze historical data to improve efficiency, reduce costs, and enhance the speed of construction.

8. **Smart Neighborhoods by Google's Sidewalk Labs (Various Locations):**

   o **AI Application:** AI is applied for neighborhood planning and infrastructure optimization.

   Sidewalk Labs, in collaboration with developers, is working on creating smart neighborhoods that leverage AI for efficient waste management, traffic flow optimization, and enhanced community services.

9. **YITU's AI City (Chengdu, China):**

   o **AI Application:** AI is utilized for smart city development, including residential and commercial areas.

   YITU Technology is involved in developing AI-driven solutions for smart city applications, including facial recognition for security, traffic management, and optimizing public services.

10. **BIM and AI Integration (Various Projects):**

- o **AI Application:** AI is integrated with Building Information Modeling (BIM) for enhanced design and construction.

- o Various residential and commercial developments leverage AI to enhance the capabilities of BIM. This integration improves design accuracy, project visualization, and collaboration among stakeholders.

These case studies showcase the diverse applications of AI in residential and commercial developments, ranging from smart city planning and energy efficiency to construction site safety and building management. As the field continues to evolve, more innovative AI-driven projects are likely to emerge, transforming the way residential and commercial spaces are designed, built, and managed.

## 16.3 Rehabilitation and Retrofitting

Rehabilitation and retrofitting projects aim to upgrade existing structures, improving their functionality, safety, and sustainability.

AI-driven technologies play a crucial role in optimizing the rehabilitation and retrofitting processes. Here are some case studies that illustrate the application of AI in rehabilitation and retrofitting projects:

1. **AI for Structural Health Monitoring - Smart Retrofitting:**
   - **AI Application:** Structural health monitoring using AI for identifying vulnerabilities and recommending retrofitting strategies.

Various projects have applied AI algorithms to continuously monitor the health of structures. These systems can analyze sensor data to detect early signs of structural deterioration and suggest retrofitting measures to enhance the building's resilience.

2. **Energy-Efficient Retrofitting with AI - The Crystal (London, UK):**

   o **AI Application:** AI is employed for optimizing energy-efficient retrofitting solutions.

   The Crystal, a sustainable cities initiative in London, utilized AI to retrofit an existing building for improved energy efficiency. AI algorithms analyzed data on energy consumption patterns and recommended retrofitting measures, leading to significant energy savings.

3. **AI for HVAC Retrofitting - Empire State Building (New York, USA):**

   o **AI Application:** AI is used for optimizing Heating, Ventilation, and Air Conditioning (HVAC) systems during retrofitting.

   The Empire State Building retrofitting project incorporated AI to optimize HVAC systems. Machine learning algorithms analyzed weather data, occupancy patterns, and other variables to adjust HVAC settings dynamically, resulting in energy savings and improved comfort.

4. **Predictive Maintenance for Retrofitting - Historic Bridges (Various Locations):**

   o **AI Application:** AI is employed for predictive maintenance to guide retrofitting efforts in historic bridges.

   In retrofitting historic bridges, AI algorithms analyze sensor data to predict potential maintenance needs. This enables a proactive approach to retrofitting, preserving the structural integrity of these landmarks.

5. **AI-Enhanced Façade Retrofitting - The Edge (Amsterdam, Netherlands):**

o **AI Application:** AI-driven solutions for optimizing façade retrofitting for energy efficiency.

The Edge building in Amsterdam utilized AI to retrofit its façade. The technology adjusted window blinds and lighting based on external conditions, optimizing natural light usage and reducing energy consumption.

6. **AI for Seismic Retrofitting - Various Projects in Seismic Zones:**

o **AI Application:** AI algorithms for seismic risk assessment and retrofitting recommendations.

In regions prone to seismic activity, AI is applied to assess the seismic risk of existing structures. Machine learning models analyze historical data and structural conditions to recommend retrofitting measures for improved seismic resilience.

7. **Smart Lighting Retrofitting - Various Office Buildings:**

o **AI Application:** AI-driven solutions for optimizing lighting systems during retrofitting.

Office buildings have employed AI to retrofit lighting systems. Smart lighting solutions use occupancy data and natural light conditions to dynamically adjust lighting levels, enhancing energy efficiency.

8. **AI for Accessibility Retrofitting - Various Public Spaces:**

o **AI Application:** AI technologies for assessing and recommending accessibility retrofitting measures.

AI is used to assess the accessibility of public spaces for individuals with disabilities. The technology analyzes spatial data and recommends retrofitting solutions to improve accessibility.

9. **Dynamic Building Envelope Retrofitting - Various Residential Projects:**

o **AI Application:** AI-driven solutions for retrofitting building envelopes dynamically.

Residential projects have applied AI to retrofit building envelopes. Dynamic solutions adjust insulation and shading based on real-time weather conditions, contributing to energy efficiency and occupant comfort.

## 10. AI for Water Management Retrofitting - Various Urban Projects:

○ **AI Application:** AI is used for optimizing water management systems during retrofitting.

Urban retrofitting projects incorporate AI to enhance water management. Machine learning algorithms analyze water consumption patterns and recommend retrofitting measures for sustainable water use.

These case studies demonstrate the versatility of AI in rehabilitation and retrofitting projects, offering solutions that range from structural health monitoring to energy efficiency enhancements. As technology continues to advance, AI-driven approaches are likely to play an increasingly vital role in optimizing the retrofitting of existing structures for improved performance and sustainability.

## 16.4 International Collaborations

International collaborations in AI-driven construction projects often involve partnerships between organizations, researchers, and industry players from different countries.

These collaborations aim to leverage diverse expertise, share knowledge, and address global challenges in the construction industry. Here are some case studies that highlight international collaborations in AI-driven construction projects:

1. **EU Horizon 2020 Project - BIM-SPEED (Building Information Modeling - Sustainably Precast Elements for Energy-efficient Buildings):**

- o **Collaboration:** Collaborative effort involving partners from several European countries.

- o **Objective:** The project focuses on advancing Building Information Modeling (BIM) and AI for the design and production of sustainable precast building elements. It involves partners from academia, industry, and research institutions across Europe.

2. **Japan-Singapore Collaboration - Smart Construction Digitalization (SCD):**

- o **Collaboration:** Joint collaboration between Japanese and Singaporean construction and technology firms.

- o **Objective:** The collaboration aims to enhance construction processes through digitalization and AI technologies. It involves implementing AI for construction site monitoring, project management, and the integration of smart technologies in construction practices.

3. **Australia-China Collaboration - AI in Construction Safety:**

- o **Collaboration:** Joint research collaboration between Australian and Chinese institutions.

- o **Objective:** The collaboration focuses on applying AI technologies to improve construction site safety. AI-driven systems are developed to monitor and analyze safety conditions on construction sites, with the goal of reducing accidents and improving overall safety standards.

4. **UK-Germany Collaboration - Autonomous Construction Vehicles:**

- o **Collaboration:** Collaboration between UK and German companies and research institutions.

- o **Objective:** The project explores the use of AI in autonomous construction vehicles. AI algorithms are developed to enable construction machinery to operate autonomously, improving efficiency and safety on construction sites.

5. **Canada-India Collaboration - Infrastructure Monitoring and Management:**

o **Collaboration:** Collaboration between Canadian and Indian engineering and technology firms.

o **Objective:** The collaboration focuses on AI-based infrastructure monitoring and management. AI algorithms are employed to monitor the condition of bridges, roads, and other infrastructure elements, with the aim of enhancing maintenance and minimizing risks.

6. **USA-South Korea Collaboration - AI for Smart Cities:**

o Collaboration: Joint collaboration between U.S. and South Korean researchers and technology companies.

o Objective: The collaboration aims to implement AI technologies in the development of smart cities. AI is applied to optimize urban planning, traffic management, and construction processes in line with the principles of sustainable and intelligent urban development.

7. **China-UK Collaboration - Digital Twin Technology for Construction:**

o Collaboration: Collaboration between Chinese and UK-based companies and research institutions.

o Objective: The collaboration explores the use of digital twin technology in construction projects, integrating AI for real-time monitoring and simulation. The goal is to enhance project efficiency and decision-making through a collaborative digital representation of the construction process.

8. **Norway-Singapore Collaboration - Offshore Construction Automation:**

o Collaboration: Joint effort involving Norwegian and Singaporean companies and research entities.

o Objective: The collaboration focuses on automating offshore construction processes using AI. Robotics and AI algorithms are developed to optimize construction workflows in challenging offshore environments.

9. **France-India Collaboration - AI for Sustainable Construction Practices:**

- o Collaboration: Collaboration between French and Indian construction and technology firms.

- o Objective: The collaboration aims to incorporate AI into sustainable construction practices. AI is applied to optimize resource utilization, energy efficiency, and waste reduction in construction projects.

**10. Netherlands-United Arab Emirates Collaboration - Smart Urban Infrastructure:**

- o Collaboration: Collaboration between Dutch and UAE-based organizations.

- o Objective: The collaboration focuses on developing smart urban infrastructure using AI. AI technologies are applied to enhance the efficiency of urban infrastructure systems, including transportation, energy, and construction.

These case studies illustrate how international collaborations in AI-driven construction projects bring together expertise from different regions to address common challenges and advance the use of AI in the construction industry on a global scale. The collaborative efforts aim to drive innovation, share best practices, and contribute to the development of sustainable and technologically advanced construction practices worldwide.

## 16.5 Lessons from Successful Projects

Several lessons can be drawn from successful AI-driven construction projects.

These lessons highlight key factors that contribute to the effectiveness and positive outcomes of such initiatives. Here are some essential lessons learned from successful AI-driven construction projects:

1. **Clear Objectives and Use Cases:**
   
   o   Lesson: Clearly define project objectives and use cases for AI applications in construction.
   
   Successful projects start with a clear understanding of the specific problems AI is meant to address. Identifying use cases

that align with project goals ensures focused and impactful implementations.

2. **Collaboration and Multidisciplinary Teams:**

   o Lesson: Foster collaboration among multidisciplinary teams.

   Successful projects often involve collaboration between construction professionals, data scientists, AI specialists, and domain experts. A diverse team brings varied perspectives and expertise, contributing to comprehensive solutions.

3. **Data Quality and Accessibility:**

   o Lesson: Ensure high-quality and accessible data.

   The success of AI models depends on the availability of relevant and high-quality data. Projects should prioritize data cleanliness, consistency, and accessibility to train accurate and effective AI algorithms.

4. **Pilot Testing and Iterative Development:**

   o Lesson: Conduct pilot tests and embrace iterative development.

   Implementing AI in construction projects is often complex. Piloting small-scale projects allows for testing and refining AI solutions in real-world scenarios. Iterative development accommodates adjustments based on feedback and evolving project requirements.

5. **Integration with Existing Systems:**

   o Lesson: Ensure seamless integration with existing construction systems.

   Successful projects integrate AI solutions into existing workflows and systems. Compatibility with established processes helps in smoother adoption and ensures that AI enhances, rather than disrupts, the overall construction process.

6. **Stakeholder Engagement and Communication:**

   o Lesson: Engage stakeholders and communicate effectively.

   : Projects benefit from involving key stakeholders early on and maintaining transparent communication throughout.

Understanding user needs, addressing concerns, and obtaining feedback contribute to successful project outcomes.

7. **User Training and Adoption Strategies:**

   o Lesson: Prioritize user training and adoption strategies.

   Successful AI implementations involve thorough training for end-users. Ensuring that construction professionals are comfortable with and understand the AI tools enhances adoption rates and overall project success.

8. **Scalability and Future-Proofing:**

   o Lesson: Design solutions with scalability and future-proofing in mind.

   Successful projects plan for scalability and account for future advancements. Scalable solutions can grow with project requirements, and future-proofing anticipates technological advancements and evolving industry standards.

9. **Ethical Considerations and Bias Mitigation:**

   o Lesson: Address ethical considerations and mitigate biases.

   Ethical considerations, including privacy and bias, are critical. Successful projects prioritize responsible AI practices, incorporating measures to protect privacy, ensure fairness, and mitigate biases in AI algorithms.

10. **Continuous Monitoring and Maintenance:**

    o Lesson: Implement continuous monitoring and maintenance.

    Post-implementation, successful projects maintain vigilance. Continuous monitoring of AI models ensures ongoing accuracy, and proactive maintenance addresses issues promptly, contributing to sustained success.

11. **Regulatory Compliance:**

    o Lesson: Ensure compliance with regulatory standards.

    Successful projects adhere to relevant regulations and standards. Compliance with data protection, privacy, and industry regulations is crucial for the long-term success and acceptance of AI-driven construction projects.

## 12. Return on Investment (ROI) Assessment:

o   Lesson: Evaluate and demonstrate ROI.

Successful projects assess and demonstrate the return on investment. This involves quantifying the impact of AI on project outcomes, such as cost savings, time efficiency, and overall project success.

These lessons highlight the importance of a holistic and strategic approach to implementing AI in construction projects. Embracing these principles can contribute to the successful adoption and integration of AI technologies in the construction industry.